New Modes of Governance

Developing an Integrated Policy Approach to Science, Technology, Risk and the Environment

D0068761

Edited by

CATHERINE LYALL and JOYCE TAIT
University of Edinburgh, UK

ASHGATE

Published by
Ashgate Publishing Limited
Gower House
Croft Road
Aldershot
Hants GU11 3HR
England

Ashgate Publishing Company
Suite 420
101 Cherry Street
Burlington, VT 05401-4405
USA

Ashgate website: http://www.ashgate.com

British Library Cataloguing in Publication Data
New modes of governance : developing an integrated policy
    approach to science, technology, risk and the environment
    1. Science and state - Great Britain 2. Technology and state
    - Great Britain 3. Environmental policy - Great Britain
    4. Risk - Government policy - Great Britain
    I. Lyall, Catherine II. Tait, Joyce
    338.9'41

Library of Congress Cataloging-in-Publication Data
New modes of governance : developing an integrated policy approach to science, technology, risk and the environment / [edited] by Catherine Lyall and Joyce Tait.
        p. cm.
    Includes bibliographical references and index.
    ISBN 0-7546-4164-3
  1. Science and state--Great Britain. 2. Technology and state--Great Britain. 3. Environmental policy--Great Britain.
    [DNLM: 1. Public Health Administration--methods--Great Britain. 2. Biotechnology--standards--Great Britain. 3. Environment--Great Britain. 4. Internationality--Great Britain. 5. Public Policy--Great Britain. 6. Research--methods--Great Britain. 7. Risk management--Great Britain. 8. Systems Integration--Great Britain. WA 540 FA1 N532 2005] I. Lyall, Catherine. II. Tait, Joyce.

    Q127.G4N49 2004
    338.941'06--dc22

                                                        2004025348
ISBN 0 7546 4164 3

Printed in Great Britain by Antony Rowe Ltd, Chippenham, Wiltshire

NEW MODES OF GOVERNANCE

Contents

PART I: NEW APPROACHES TO GOVERNANCE

PART II: DEVELOPING AN INTEGRATED POLICY APPROACH

PART III: THE LIMITS TO INTEGRATION

List of Tables

List of Figures

List of Contributors

John Adams, Professor of Geography at University College London, was a member of the original board of directors of Friends of the Earth in the early 1970s and has been a participant in debates about transport planning and the management of environmental risks ever since. He has published widely on these themes in both specialist journals and the national press. He is also a frequent contributor to radio and television programmes on related controversies.

Frans Berkhout is Director of the Institute for Environmental Studies (IVM) at the Vrije Universiteit Amsterdam. Frans has extensive research and research management experience. His recent work has been concerned with technology, policy and sustainability, with special emphasis on the links between technological innovation and environmental performance in firms, the measurement of sustainability performance, futures scenarios studies, business adaptation to environmental change, and policy frameworks for innovation and the environment.

Joanna Chataway is a Senior Lecturer in the Development Policy and Planning Department at the Open University. She is also Principal Investigator on two projects core to the ESRC Innogen Centre and on another research project associated with the centre. These projects are all concerned with North/South knowledge and technology flows in genomics and biotechnology and have a particular focus on public-private partnerships. Her research interests also include the dynamic between policy and company strategy in biotechnology and agri-chemicals.

Julia Hertin is a Research Fellow at SPRU – Science and Technology Policy Research at the University of Sussex. She is a political scientist and her main research interests are the integration of sustainability issues into public and private decision-making and understanding the determinants of environmental improvement in business.

Catherine Lyall is a Research Fellow at the University of Edinburgh where she is studying the effects of policy and regulatory regimes on developments in genomics. Her other research interests span policy issues related to the governance of science, technology, innovation and knowledge systems; research evaluation and the management of interdisciplinary research, and she is currently completing a part-time PhD exploring the impact of devolution on the policy-making process for science and innovation in Scotland.

James McQuaid has a background in engineering and has spent most of his career in research and policy development in government. He was Chief Scientist of the UK Health and Safety Executive from 1992 until his retirement in 1999. During that period he was also chairman of the Interdepartmental Liaison Group on Risk Assessment. The Group had as its remit to help secure coherence and consistency within and between policy and practice in risk assessment as undertaken by government, and to help disseminate and advance good practice. He is now a Visiting Professor in the School of the Built Environment, University of Ulster and has similar appointments at the University of Sheffield and Queen's University Belfast.

Joseph Murphy is Assistant Director of the ESRC Genomics Research Forum at the University of Edinburgh. He is an interdisciplinary social scientist with a particular interest in environmental policy and politics. In the past he has done work on the relationship between environmental regulation and industrial innovation and sustainable consumption as a public policy problem. He recently completed a two-year ESRC project examining the conflict between the European Union and the United States over genetically modified organisms and its impact on regulatory standards. He is also currently running an ESRC network on the governance of sustainable technologies.

Perri 6 is Professor of Social Policy at Nottingham Trent University. He worked previously at the University of Birmingham, King's College, Strathclyde University and the think tank, Demos. He has written widely on public policy. His most recent books include "E-governance" (2004) and "Towards holistic governance", and among his recent articles are studies on consumer choice in public services, identity cards, data sharing and privacy, joined-up government cross-nationally, New Labour's public management reforms, risk perception and the sociology of knowledge and a neo-Durkheimian theory of institutional viability.

Thomas Reiss has a PhD in Molecular Biology from Freiburg University. After postdoctoral research at the DOE Plant Research Lab, Michigan State University, USA, he joined Fraunhofer ISI in 1987 where he has been Head of the Innovations in Biotechnology Department since 1996. His current research focuses on national and sectoral innovation systems, industrial innovation, and the evaluation of innovation policies in biotechnology. He has carried out several projects on policy issues related to biotechnology for the European Commission and was co-ordinator of the EPOHITE project. Since 2002 he has been a member of the OECD TIP focus group on pharmaceutical biotechnology.

Frank Rennie is the Head of Research and Postgraduate Studies at Lews Castle College and Course Leader of the MSc in Managing Sustainable Rural Development at the UHI Millennium Institute in the Highlands and Islands of Scotland. His research interests lie in the general areas of rural and community development, especially in community-based approaches to integrated sustainable development. Recent work has been on new approaches to online education and blended learning on and in rural communities. He is an advisor to several government programmes and committees and is a Fellow of a number of learned societies.

Graham Spinardi is a Senior Research Fellow in the Research Centre for Social Sciences at the University of Edinburgh and a Lecturer in the University's Science Studies Unit where he convenes courses on Social and Economic Perspectives on Technology and the Politics of Science and Technology. His research interests encompass military technology, electronic data interchange, environmental innovation, and the exploitation of inventions.

Joyce Tait is a Professorial Fellow and Director of the ESRC Innogen Centre at the University of Edinburgh. She has an interdisciplinary background covering both natural and social sciences. She has specialised in systemic approaches to complex issues in the contexts of innovation and environmental management in: risk assessment and regulation; policy analysis; technology management in the agrochemical and biotechnology industries; sustainable development; public attitudes and communication; strategic and operational decision-making in companies and public bodies; land use and crop protection.

Robin Williams is Director of the Research Centre for Social Sciences/Institute for the Study of Science, Technology and Innovation, and a Co-Director of the ESRC Innogen Centre at the University of Edinburgh. After doctoral research into the control of industrial hazards, he has undertaken extensive research into technology and work organisation, in relation to the design and implementation of integrated corporate information systems and inter-organisational networks. He has developed an international programme of interdisciplinary research into the social shaping of technology. His more recent work has focused on emerging technologies including wireless technologies, biotechnology and nanotechnology.

Preface and Acknowledgements

This book is the culmination of a number of activities and brings together a range of collaborators with whom we have worked on a variety of projects. A grant under the UK ESRC's research seminars competition in 2000 enabled us to explore the theme of *Best Practice in Evidence-Based Policy in a Devolved Context* and prompted us to think about the development of new approaches to policy for science, technology and the environment in the context of the devolution of powers to the Scottish Parliament; the need for Scottish institutions to contribute effectively to debates at the UK and EU levels; and the desire to contribute to the wider international policy revolution.

These topics were further developed through three subsequent series of training seminars for policy-makers and by the establishment of the ESRC Centre for Social and Economic Research on Innovation in Genomics (the Innogen Centre) which has enabled us to develop these themes specifically in the area of the life sciences. Our own programmes of research within the ESRC Innogen Centre are situated within both national and international policy contexts, strengthened by our engagement with European policy-makers and interactions with other European colleagues. Many of the authors in this book also adopt this perspective and are able to offer insights into international policies for science, technology and innovation.

A common thread running through all of these activities, and consequently through the chapters which follow, has been a desire to explore and improve the quality of decision-making by developing integrated, interdisciplinary approaches to policy research and advice and engendering a better appreciation of the role of science in the policy-making process.

We are indebted to all of our collaborators for their support and patience throughout the production of this book and would particularly like to thank our colleagues, Eileen Mothersole and Moyra Forrest, for their assistance with the preparation of the manuscript and index respectively.

Catherine Lyall and Joyce Tait

PART I
NEW APPROACHES TO
GOVERNANCE

Chapter 1

Shifting Policy Debates and the Implications for Governance

Catherine Lyall and Joyce Tait

Policy-making is in a state of flux and governments are stressing the need for more integrated or "joined up" policies to deal with the complex issues now facing society. There is an ongoing policy revolution taking place at global, European, UK and regional levels but the buzz-words that are often used to describe this revolution – "Third Way", "joined-up policies", "what works" – are sometimes regarded with scepticism.

The *Modernising Government* initiative (HM Government, 1999) in the UK had as one of its primary themes the need for forward looking, more integrated policy-making. The policy environment for science, technology and innovation (STI) is one of the areas most in need of an integrated policy approach. Getting this right is vital to nations' economic prosperity and hence their ability to deliver on other social and environmental policies. However, there is arguably less integration in this area than in many others and, so far, there has been little guidance on what integrated policies would consist of and how they might be delivered.

The Governance Perspective

Theories of governance draw on a range of disciplinary perspectives and operate at multiple levels (local, regional, national and supra-national). In the most common current usage of the term, "Governance" is seen as implying a move away from the previous *government* approach (a top-down legislative approach which attempts to regulate the behaviour of people and institutions in quite detailed and compartmentalised ways) to *governance* (which attempts to set the parameters of the system within which people and institutions behave so that self-regulation achieves the desired outcomes), or put more simply, the replacement of traditional "powers over" with contextual "powers to" (Pierre and Peters, 2000). In such a governance system, permeable and flexible system boundaries will facilitate communication and will support the achievement of higher level goals. These assumptions underline the switch from *government* to *governance* in debates about the modernisation of policy systems implying a switch from constraining to enabling types of policy or regulation (i.e. from "sticks" to "carrots"). The UK literature tends to discuss the emergence of "new" modes of governance, and in particular "joined-up government", in terms of a response to Thatcherite reforms, specifically as a way of coping with the fragmentation effects of New Public Management (NPM) (Newman, 2001; Bevir and Rhodes, 2003) although its antecedents arguably go back much further. The reconnection in the 1990s of the work on networks in policy-making with studies in organisational sociology of networks of service provision provided a spur to the rethinking of governance in network terms which in many ways has provided a critical link to the joined-up government debate (Perri 6, 2004).

The governance debate has its origins in various disciplines spanning institutional economics, international relations, organisational studies, development studies, political science, and public administration (Stoker, 1998). As others have noted, the term "governance" is currently applied to "everything from corporations to rural society" (Sloat, 2002) and the academic literature on the subject of governance has been described as "eclectic and relatively disjointed" (Stoker, 1998). Its meaning is contested and often lacks definitional clarity (Bache, 2003) although most commentators accept that "governance" is no longer a synonym for "government". However, Stoker (1998) argues that, as governance is ultimately concerned with creating the conditions for ordered rule and collective action, its outputs are no different from those of government. What is significant is the difference in processes.

Most would accept that governance refers essentially to the increased role of non-government actors in policy-making (Bache, 2003) and it is generally regarded as implying an increasingly complex set of state-society relationships in which networks rather than hierarchies dominate the policy-making process (Bache, 2003). The current work adopts the view that "governance" refers to the development of governing styles in which boundaries between and within public and private sectors have become blurred; in this approach the role of the state changes from being the main provider of policy to one of facilitating interaction among various interests (Sloat, 2002). In this context, government's role is increasingly one of co-ordination and steering (Bache, 2003).

From a governance perspective, the process of governing is an interactive one because no single actor has the knowledge and resource capacity to tackle problems unilaterally (Kooiman, 1993) and the powers of government tiers are no longer clearly distributed, as co-operation replaces hierarchy and legislative competences are shared among several levels (Sloat, 2002). In summary, this perspective focuses on the co-ordination of multiple actors and institutions to debate, define and achieve policy goals in complex political arenas such that the state no longer dominates the policy-making process and decisions are made by "problem solving rather than bargaining" (Sloat, 2002).

"Governance" is therefore a term that takes different meanings in the hands of different authors and Newman (2001) suggests that, in fact, it comprises multiple and conflicting strands and is constituted by disparate forms of power. Some authors take a "state centric image of governance" (Pierre and Peters, 2000) in contrast to Rhodes' approach (Rhodes, 1997), which downplays the role of central government. Others focus on the complexity, dynamics and diversity of interactive social-political governance where the state still has a role in steering society (Kooiman, 1993). In this scenario there has been a shift from formal powers to political capabilities so that there is now less reliance on coercive policy instruments and a greater reliance on more subtle techniques. This has led to a restructuring of state institutions, creating agencies, quangos and other institutional forms that operate at considerable distance from control by the political elite (Pierre and Peters, 2000). Nevertheless, these authors contend that states as centres of governance still play a defining role in the economy, in international relations, and in many areas of domestic politics and policy. This perspective highlights concerted public-private efforts and co-operative rather than adversarial policy strategies. Pursuing this collective interest through different forms of governance on and between different institutional levels requires closer, more continuous and more informal contacts between political institutions and their environment (Pierre and Peters, 2000 p.196).

From an academic standpoint, the governance perspective might just be "a simplifying lens to a complex reality" (Stoker, 1998) but its practical value rests in its capacity to provide a framework for understanding changing processes of governing, characterised by processes of adaptation, learning and experiment (Stoker, 1998). This approach necessitates a new policy design where the state acts as a moderator and enabler within a network-oriented polity rather than a hierarchical interventionist approach. This requires a "new mode of governance" with, on the one hand, more elaborate forms of institutionalised co-ordination between the European level and national and regional levels, and on the other hand, a continuous reflection on appropriate principles of governance with respect to both strong leadership and participatory approaches which in turn requires a re-organisation of policy administrations in a way that enables flexible, horizontal co-ordination and exchange (Edler et al., 2002).

Policy Integration

In general, public policy is not used to tackling problems in an integrated manner and, in particular, policy approaches for areas such as innovation are usually compartmentalised in different departments and agencies that compete for power rather than co-operate to tackle policy issues (Cooke et al., 2000 p.142). Despite current and emerging socio-economic and political developments where policy-makers in most industrialised countries are trying to reform traditional policy approaches, intradepartmental rather than overall co-ordination would still seem to be the preferred policy mechanism in the UK at least where individual agencies are allowed self-determination while trying to avoid undue duplication of effort or pursuit of conflicting goals in different parts of government (Ronayne, 1984 p.141). In practice, it is still true that co-ordination in Britain tends to mean cross-membership of committees: an "insider's world" where a relatively small group of senior civil servants, elite scientists, and influential industrialists move from committee to committee (Ince, 1986) and co-ordination between policy domains remains the exception rather than the rule.

Over the past thirty years there has been a steady shift in the emphasis of research policies at national and European levels, to obtain better value for money from public investment in research by ensuring that both curiosity-driven, fundamental research and applied research contribute as much as possible to improving competitiveness at national, European and international levels. New approaches to governance are being developed under a variety of labels at different institutional levels in many European countries. In the UK, what some see as an on-going policy revolution and others regard as a more managerial response to the often unintended consequences of NPM (Bevir and Rhodes, 2003), has been referred to as the "Third Way", which promotes a network-based polity as an alternative to bureaucracies and markets (Bevir and Rhodes, 2003; Giddens, 1998)[1], with a strong commitment to more integrated or "joined-up" approaches to policy. Others have advocated "holistic governance" (6 et al., 2002) which goes beyond simply stitching together the plethora of government committees and policy documents and instead takes a more grassroots approach which moves away from a model of government that is structured around functions and services and instead focuses on solving "wicked" problems (6, 1997).

Some parameters of the new governance-based policy-making systems are relevant to STI policy, such as initiatives on policy integration, evidence-based policy, the use of standards and guidelines linked to policy evaluation, encouragement of openness, stakeholder involvement and consultation, and avoidance of unnecessary regulatory burdens. However, despite frequent references to the need for more integrated approaches to policy development, new governance initiatives in the UK are largely socially-oriented and ignore STI-related issues (Lyall and Tait, 2004).

The development of new governance structures, for example in the UK under the *Modernising Government* agenda (HM Government, 1999) focuses on modernising the processes of government, including a framework for excellence in policy-making and a strong emphasis on learning lessons from policy experience in other countries. The over-arching ethos is "what matters is what works" (Davies et al., 2000) with, at least in theory, a much freer flow of ideas across governments and government departments and from one level of government to another, focusing on ideas that can contribute to an effective system of governance, rather than on the ideology that generated the ideas.

The *cri de coeur* of the current Labour Government for "joined-up" policy is reflected in the goal of the *Modernising Government* initiative to develop a more integrated approach to policy-making, and a series of Cabinet Office publications (for example, Cabinet Office Performance Innovation Unit, 2000; Cabinet Office

---

[1] The term "Third Way" seems to have been dropped from the political lexicon in the UK recently, although the new governance approaches it described are continuing to be developed.

Strategic Policy Making Team, 1999) aims to improve policy formulation and implementation in areas that cut across the policy boundaries of traditional government departments.

Effective policy integration would imply that science-related policies ought to be crucial components of new governance initiatives but we find little evidence of their inclusion. The *Modernising Government* agenda concentrates almost entirely on the social policy arena covering social welfare, crime, health and education, these being the areas which focus groups tell government ministers are of most concern to voters. Science, technology, and innovation are apparently of lesser concern to voters as they are not linked in the public mind (and hence less likely to be linked by governments) to national competitiveness which generates the wealth to support the other functions, although others would argue (6, 2004) that there *have* been attempts with, for example, the publication of the Competitiveness White Paper (DTI, 1998) which aspired to provide an integrated framework of tax, subsidy, regulation, trade, patent and regional policy for the development of science-based industries in the UK, particularly in respect of the horizontal co-ordination of agencies' priorities.

As Tait et al. (2004) note, a consistent theme throughout this new governance agenda is the need for more integrated or "joined-up" policy approaches to remove contradictions, inconsistencies and inefficiencies caused when policies or regulations emerging from different government departments or different levels of government (regional, national, international) contradict one another or provide incompatible signals to policy targets. Policy integration is also needed to deal with the complexity and uncertainty associated with many decisions concerning science and technology. However, integration has itself become more difficult as the diversity and policy competence of interested stakeholders and publics has increased.

European Governance

Similar trends are also beginning to emerge at the EU level with important documents being published on European governance, the European Research Area (ERA), and developments in the Sixth Framework Programme (FP6). As in the UK, there is evidence of difficulty in integrating policies and particularly in spanning the divide between science/technology and society.

The gap between innovative thinking on governance in general and developments in science and technology-related policies is also apparent at the EU level. The *White Paper on European Governance* (Commission of the European Communities, 2001) has only one brief reference to the word "science" in the context of managing "...the challenges, risks and ethical questions thrown up by science and technology". There are no references to "evidence", and for "research" there is one mention of "research centres" and one to the ERA, although there are references to scientific committees and the need for their advice to be made publicly available. The overall impression is that science-related issues are of only peripheral interest in the context of European governance although they presumably come into the picture downstream, as a part of policy implementation in sectoral documents such as those concerned with telecommunications or human embryology, rather than being integrated at a high level into the overall governance and policy development process.

The document on the ERA is the main focus of innovative EU thinking on science and research-related issues. One of its main policy planks is the forging of closer links between the EU Framework Research Programmes and the research systems of EU member states. The ERA will be implemented partly through FP6, involving also major changes in the organisation of research in Europe. Prior to the development of ideas on the ERA, and influenced to some extent by UK thinking on the development of Foresight, the Fifth Framework Programme (FP5) took a new direction by giving a strong emphasis to interdisciplinary integration, particularly between the natural and social sciences. FP5 targeted Key Actions to socio-economic needs and guided research collaboration among EU nations in a manner that increasingly included socio-economic components. It would be unrealistic to expect such a major change in research orientation and management to bear fruit within the time scale of a single programme and it is unfortunate that FP6 has largely abandoned the innovative approach on interdisciplinary research pioneered by FP5. Although essential if Europe is to compete effectively in a global economy, integrative approaches challenge many vested interests in both academic and policy spheres and, as we have noted, there have been strong reactions against it from several directions.

Global Governance Issues

As outlined by Tait and Bruce (2004), the increasingly rapid pace of technological innovation and the increasing size and power of multinational companies are leading to globalisation of production and trading systems accompanied by pressures for further trade liberalisation.

The emerging system of global governance is being mediated through international organisations like the World Trade Organization. However, such changes diminish the sense of power and influence of individual citizens and appear to negate local and national democratic processes, raising fundamental questions of sovereignty and governance at national and regional levels. They are also being opposed by increasingly vocal and well organised public groups acting against globalisation and the pressures that are driving it. In the context of developments in genetically modified crops, Tait and Bruce (2004) referred to the internationally organised consumer boycott as "a new instrument of global governance".

Giddens (1999) noted this tension between pro- and anti-globalisation forces. He referred on the one hand to "… the mobilising dynamic of a society bent on change, that wants to determine its own future…", and on the other hand he noted that we now live in a world where innovation and technological change have generated hazards that are regarded as more threatening than so-called natural hazards.

National science-related policies can no longer operate effectively without considering the pressures and constraints imposed at the global level. These include:

- international trading relationships;
- intellectual property rights;
- the relevance of regionally-based technology clusters in the context of modern information and communication networks;
- public support for, or opposition to, individual innovations.

Horizontal integration of issues, polices and initiatives is in some cases lacking altogether. Where it has been attempted it is being modified because of difficulties in its implementation. For example, at EU, UK and regional levels, there is so far a general lack of integration between the modern approaches to governance being developed in the social policy arena and policies for science, technology and innovation which still seem to be driven by an old-fashioned, linear conception of innovation systems (Tait and Williams, 1999). However, in the chapter which follows, Perri 6 will argue that these concerns about a lack of horizontal co-ordination mechanisms at a global level are overstated.

Actors and Outcomes

Demands for more integrated approaches are driven by the increasing realisation that policies often deliver much less than is expected or intended, because of counter-productive interactions among the key actors, or because the policies arising from different sectors of the policy environment conflict with one another. On the other hand, where interactions among the actors or the policies are supportive, the desired outputs can be achieved more rapidly and at less public cost.

One strand of our research on governance (Tait et al., 2004) focuses on the key actors who are the targets of policies. These actors are linked to one another in often-complex webs of interaction and are influenced by a policy environment which includes regulations, fiscal measures, manipulation of the infrastructure, and a range of non-statutory, voluntary incentives and constraints. The overall aim of this mix of policies is to encourage the delivery by these actors of a set of outputs that are regarded as socially desirable.

The chapters in this book describe the perspectives of the policy-makers themselves and how they are attempting to influence actor/stakeholder networks in different policy areas and in different industry sectors. The range of relevant policies varies from one sector to another. Policies also differ in their nature (from legally enforceable regulatory instruments and fiscal policies, through voluntary codes of conduct, to government initiatives like Foresight) and are directed to different targets in the actor network. It is part of the policy-maker's stock in trade to match the policy mix in their area of influence to the intended purpose (the outputs to be delivered). But real world complexity usually means that the match is less than perfect.

Different Modes of Policy Integration

Our analysis covers two radically different types of policy integration:

Vertical Integration

Policy integration across levels of governance depends mainly on the ability to communicate effectively across system boundaries and the institutional structures determined by government policy-making at the highest levels has a major influence on the effectiveness of such communications. Vertical integration is thus mainly a function of the institutional structures determined by policy-making at senior government levels and its most important constituents are effectiveness of communications across levels of government. Ideally, vertical communication across these boundaries should be a two-way process, leading to accommodation by higher levels to the needs of the lower levels as well as the reverse process. However, there is a potential conflict between exercising discretion and hierarchical control and this gives rise to questions about whether every layer within any given hierarchy has to be vertically integrated. Is hands-off management, the delegation of responsibilities and target-setting the right way to handle a complex system? If the purpose of vertical integration is to ensure that actions at every level are consistent with policies decided at the top level, what happens when things go wrong? Effective vertical integration often implies top-down control with some form of sanction imposed where higher level policies are ignored or flouted so that often the reaction is to restore the controls rather than examine faults at the top level.

Horizontal Integration

Horizontal integration takes place across departmental boundaries, for example the ideal, but so far patchy, integration between science and technology policy and social and environmental policies in the UK in the development of new approaches to governance. Integration in this case poses similar challenges to that of interdisciplinary research in academic organisations (Tait et al., 2002; Bruce et al., 2004).

Institutional structures are important here but they do not determine the effectiveness of integration. In the UK there have been numerous examples of amalgamation of government departments, with integration as one of the main aims, where the staff involved have continued to operate within their pre-existing boundaries, with little interaction across these old boundaries.

Communication is also important but the focus of the communication is different and it imposes different challenges. As with interdisciplinary research in academia, each policy area has its own specialist language and this leads to difficulties in effective communication across boundaries (Tait and Lyall, 2001). Likewise, career structures for public servants reward those who specialise and it is difficult to make a career by "trespassing" across traditional boundaries.

Most important, the impact of effective horizontal integration is a *loosening* of control and the introduction of greater complexity into policy implementation processes.

Horizontal policy integration, despite the importance we would attach to it, is therefore much more difficult to achieve than vertical integration. It cuts across the career structures of public servants, raises communication difficulties and lessens the ability of individual departments to exercise control in their own spheres.

The challenges exercised by horizontal policy integration and by interdisciplinary research in academia are similar in many respects, and interdisciplinary research itself has an important role in STI-related policy. To date, experience of interdisciplinary integration in FP5 has been mixed, but it is important that the EC learns from experience and adapts future programmes accordingly, rather than abandoning the experiment. Discussions with scientists who have worked in both Europe and the USA have led to the conclusion that America manages academic interdisciplinary integration much more effectively than we do in Europe, and this could be a significant component of their relative competitive advantage in many areas (Tait et al., 2002; Bruce et al., 2004).

These difficulties are related to the linearity of the assumed model of innovation. Current assumptions see "society" entering the picture as a market for the products of innovation at the end of the development pipeline, but not as a partner in their development. Innovative companies may engage in sophisticated market forecasting techniques, but they often have a very restricted understanding of what constitutes their market. Likewise, many of them fail to consider the policy environment into which their products will be launched. The current UK debate on the introduction of GM crops illustrates this point perfectly and raises questions about stakeholder engagement and the extent to which it is appropriate to involve society and at what stage (Tait, 1993; Tait et al., 2001).

Aims of the Book

Policy co-ordination may be the philosopher's stone of modern government – "ever sought but always just beyond reach" (Bevir and Rhodes, 2003), but in attempting to develop new integrated approaches to policy for science, technology, risk and the environment, this book seeks to make the following contributions to this quest:

- to improve the quality of decision-making by developing integrated, interdisciplinary approaches to policy research and advice and engendering a better appreciation of the role of science in the policy-making process;

- to contribute to the pioneering of new approaches to policy development in the context of the Europe-wide debate on regionalisation;
- to consider the issue of policy integration from a regional, national and international perspective in order to foster the development of STI;
- to help policy-makers and practitioners see and interact with the broader picture while continuing to focus on their individual areas of expertise and responsibility;
- to share best practice and provide an introduction to the policy-making process for those new to this area.

Most aspects of government are joined up to some extent but our authors will try to address, for each of the policy areas identified above, questions about where they are joined up, to what extent, and where gaps exist where further integration might be helpful. The following chapters offer a range of perspectives from UK, European and international policy regimes. Some authors question the extent to which policy integration is achievable or, indeed, desirable.

Some might argue that "globalisation" has made the governance of technological change more difficult but Perri 6's chapter which forms the second chapter in Part I argues that much depends on what expectations one has for the standard of governability. This chapter examines the ways in which technological risks, opportunities and the rise of new technology-based industries and the decline of technology-based industries are managed, first in the domestic and then in transnational contexts. Within these categories, the available tools for governance are considered, and their relative weight and interdependence explored. Perri 6 does not subscribe to the general trend toward deregulation, the decline of control and the rise of other instruments of governance, nor does he support the view that there is a general drift toward a "precautionary principle" which is being used to justify the resurgence of control methods. Instead, he suggests a much more complex picture where the tools of governance are always interdependent such that the resulting system of governance is best understood as a dynamic disequilibrium system. A central part of his argument is that we shall not understand how the governance of technology works nor how it could work differently, if we do not see the whole system: understanding better the relationship between the governance of risk, opportunity and decline is thus an important task for understanding the dynamics of governance.

Next, Spinardi and Williams draw attention to the dynamism, the serendipity and the unpredictability surrounding advances in scientific and technological knowledge and the implications that this has for policy-making. In their chapter, they argue that the turbulence of new fields of enquiry such as biotechnology and nanotechnology – so-called "breakthrough S&T" – presents new challenges for conventional science and technology policy. The innovation pathways and socio-economic outcomes of new technologies are often far removed from initial presumptions and a key challenge for research policy revolves around the assessment of the potential "coupling" between a new approach and a user requirement (either existing or to be created). Policy-makers therefore need to be attuned to the significance of paradigm shifts in science and technology as these can, for example, call into question existing criteria for assessing technology, and expectations about its social and economic outcomes and implications. These uncertainties pose enormous difficulties for the governance of breakthrough S&T, where policy is charged with the task of supporting research that will bring profound advances in understanding, as well as with promoting the exploitation of new knowledge.

The second part of the book offers a number of case studies spanning a range of policy arenas related to science, technology and innovation. First, Reiss and Tait consider the performance of European countries in promoting the life sciences, along with a range of targeted and generic policy instruments adopted to support these aims. They examine a number of Foresight studies and their potential role as a mechanism for integration of STI-related policies and argue that the more sophisticated of these Foresight initiatives could begin to develop as mechanisms to illustrate how life science innovation will play a role in determining societal futures and also how scientific, technical and social aspects of life science innovation will interact with one another to determine the future shape of these industry sectors. Reiss and Tait suggest that, in order to achieve effective, integrated governance of the life sciences, policy-makers will need to strike an appropriate balance between promotion and regulation of innovation; between what is feasible technically and commercially and what is publicly acceptable and desirable; and to deliver the claimed societal benefits without challenging accepted societal norms. A consequence of this is that policy needs to be an active moderator, bringing together key actors and technologies, and leading to an increasing focus on networking.

Risk is a prime example of an issue that pervades policies across government and for which there is a recognised need for integration of policy-making. Risk issues do not as a rule map uniquely on to the policy remits of individual government departments, which gives rise to the possibility of lack of coherence and consistency of decisions in different policy contexts. McQuaid offers a former policy-maker's perspective on developing an integrated approach to risk, and uses the experience of the UK Interdepartmental Liaison Group on Risk Assessment (ILGRA) as a case study to draw some lessons pertinent to the operation of a largely horizontal mode of integration. His chapter highlights the importance of self-adaptation which requires an awareness of differences (and the contextual reasons for them) so that individual actors can then decide how to adapt in order to ensure joined-up working.

Next, Rennie looks at the special challenges facing issues of policy and governance from a rural perspective where policy domains span a range of sustainability issues (environmental, social and economic), set against the context of rapid agri-environmental changes, debates about land use and the overlapping bureaucracy relating to environmental protection. He describes how the rural debate has moved far beyond the immediate concerns of how much food farmers can produce, to encompass fundamental questions about how we envision our society to be, how we ensure a sustainable quality of life for all citizens, and what role the rural environment plays in this bigger picture. The need to move from representative governance to more participatory structures, from policy issues that are solely top-down to those that reflect local and regional priorities within a national and international framework, will not be easy. He calls for simplification, integration, and re-prioritisation but at the same time rural development theories need to become more sophisticated, more holistic, and more balanced in their mix of quantitative and qualitative measurement of sustainable rural development for the forthcoming decades. Rennie also picks up the themes of vertical and horizontal integration, noting that the amount of re-thinking required in order to ensure both vertical integration (between central government and the peripheral regions) and horizontal integration between all policy areas that impact upon rural areas is a huge, and as yet barely addressed political challenge, but one that is central to the analysis of governance in the next few decades.

Policy-makers sometimes claim that transport policy is more "joined up" than many other policy areas. However, Adams argues that this claim usually refers to horizontal integration between infrastructure planning and transport policies and that there are no effective mechanisms, horizontal or vertical, for integrating the wider societal impacts into planning and transport policy-making. Adams suggests that connecting increasing mobility ("hypermobility") to progress continues to guide transport and communications policy. In discussing some of the downsides of the hypermobile society, Adams challenges the popular notion of governance and concludes that bottom-up "governance" – as distinct from top-down government – is not possible in a fast-moving, anonymous, low-trust, paranoid hypermobile world.

Adams' conclusion previews the theme of the third and final part of the book which begins to explore the limits to integration. Hertin and Berkhout's chapter examines the issue of environmental policy integration and its relevance for a governance approach oriented towards the development of sustainable technologies. It draws on practical experiences with environmental policy integration in EU policy-making and provides an analysis of recent initiatives such as the EU Cardiff process, the Sustainable Development Strategy and the Impact Assessment procedure, to explore the opportunities for, and barriers to, environmental policy integration. These authors demonstrate that experience at the EU level shows that environmental policy integration has proven difficult in practice and that progress is hindered by the fact that initiatives have been developed in what appears to be an unco-ordinated process, characterised by hasty changes and superficial compromises. The process of integration has not yet been pursued in a strategic and co-ordinated way and has yielded a range of well-meant, but ad hoc arrangements that are, as yet, inadequate to the task. Significantly, environmental policy integration at the EU level is dependent on better integration in member states. In offering a number of insights into lessons for policy, Hertin and Berkhout raise the possibility that policy integration could be seen by some as a convenient rhetorical position taken up at the EU level that has little substance and merely represents symbolic politics or a deliberate strategy to water-down environmental policy. They conclude that integration requires realism and that, without a dispassionate approach to the limits of integration, the whole project risks being little more than gestures and missed opportunities.

Murphy and Chataway's chapter considers some of the implications for governance and policy integration at the international level. Their chapter focuses on the international agreements and institutions that deal with the possible environmental or human health risks associated with trade in genetically modified organisms (GMOs) and shows how, with the emergence of the EU-US conflict over GMOs, these agreements and associated institutions were drawn into a complex international governance process. These authors suggest that, in some cases, lack of integration can be an outcome of governance and they argue that, in the case of GMOs, the absence of integration played an important role in the "management" of a conflict for a period of time, because of the political possibilities created by it. This chapter suggests that there are considerable risks associated with assuming that policy integration is always a good thing. From Murphy and Chataway's perspective, governance is a complex process involving the interaction of multiple stakeholders, often with different definitions of "the problem", in numerous fora at different political levels and is unlikely to be compatible, practically and theoretically, with the idea of integration.

Having considered a range of policy-relevant problems in a variety of application areas – life sciences, risk analysis, rural policy, transport, environmental policy and international trade agreements – our final chapter reflects on the reasons why new policy approaches are required. Drawing on examples from the preceding chapters, the final chapter considers both the role of evidence in policy-making and the place of stakeholder engagement in the new governance agenda and concludes by discussing the feasibility and desirability of policy integration within the new mode of governance for science, technology, environment and innovation.

References

6, P. (1997), *Holistic government*, London: Demos.

6, P. (2004), Personal communication.

6 P., Seltzer, K., Leat, D. and Stoker, G. (2002), *Towards Holistic Governance: the New Agenda in Government Reform*, Basingstoke: Palgrave.

Bache, I. (2003), 'Governing through Governance: Education Policy Control under New Labour', *Political Studies*, **51**(2), pp.300-314.

Better Regulation Task Force (2003), *Scientific Research: Innovation with Controls*, London: Cabinet Office.

Bevir, M. and Rhodes, R.A.W. (2003), *Interpreting British Governance*, London: Routledge.

Bruce, A., Lyall, C., Tait, J. and Williams, R. (2004), 'Interdisciplinary Integration in Europe: the Case of the Fifth Framework Programme', *Futures*, **36**(4), pp.457-470.

Cabinet Office Performance Innovation Unit (2000), *Wiring it Up. Whitehall's Management of Cross-cutting Policies and Services*, London: The Stationery Office.

Cabinet Office Strategic Policy Making Team (1999), *Professional Policy Making for the Twenty First Century*, London: Cabinet Office.

Commission of the European Communities (2001), *European Governance: a White Paper*, Brussels: Commission of the European Communities.

Cooke, P., Boekholt, P. and Todtling, F. (2000), The Governance of Innovation in Europe. Regional Perspectives on Global Competitiveness, London: Pinter.

Davies, H.T.O., Nutley, S.M. and Smith, P.C. (eds) (2000), *What Works? Evidence-Based Policy and Practice in Public Services*, Bristol: The Policy Press.

DTI (1998), 'Our Competitive Future: building the knowledge driven economy', White Paper Cm 4176, London: DTI.

Edler, J., Kuhlmann, S. and Smits, R. (2002), 'New Governance for Innovation. The Need for Horizontal and Systemic Policy Co-ordination' in *New Governance for Innovation? The Need for Horizontal Policy Co-ordination*, Karlsruhe: Fraunhofer ISI.

Giddens, A. (1998), *The Third Way: the Renewal of Social Democracy*, Cambridge: Polity Press.

Giddens, A. (1999), BBC Reith Lectures 1999, http://news.bbc.co.uk/hi/english/static/events/reith_99/.

HM Government (1999), *Modernising Government White Paper*, London: The Stationery Office.

Ince, M. (1986), *The Politics of British Science*, Brighton: Wheatsheaf Books.

Kooiman, J. (1993), Modern Governance. New Government-Society Interactions, London: Sage.

Lyall, C. and Tait, J. (2004), 'Foresight in a Multi-level Governance Structure: Policy Integration and Communication', *Science and Public Policy*, **31**(1), pp.27-37.

Newman, J. (2001), Modernising Governance. New Labour, Policy and Society, London: Sage.

Pierre, J. and Peters, B.G. (2000), *Governance, Politics and the State*, Basingstoke: Macmillan.

Rhodes, R.A.W. (1997), Understanding Governance. Policy Networks, Governance, Reflexivity and Accountability, Buckingham: Open University Press.

Ronayne, J. (1984), *Science in Government*, Caulfield East Victoria: Edward Arnold (Australia).

Sloat, A. (2002), 'Governance: Contested Perceptions of Civic Participation', *Scottish Affairs*, **39**(Spring), pp.103-117.

Stoker, G. (1998), 'Governance as theory: five propositions', *International Social Science Journal*, **50**(155), pp.17-28.

Tait, J. (1993), Written evidence on behalf of ESRC to Report of House of Lords Select Committee on Science and Technology on Regulation of the United Kingdom Biotechnology Industry and Global Competitiveness, 7th Report, Session 1992/93, London: HMSO.

Tait, J. and Bruce, A. (2004), 'Global Change and Transboundary Risks', in T. McDaniels and M. Small (eds), *Risk Analysis and Society: an Interdisciplinary Characterisation of the Field*, Cambridge University Press, pp 367-419. (Commissioned by Society for Risk Analysis for the International Symposium on *Risk and Governance*, Warrenton, VA, USA, June 2000.)

Tait, J., Chataway, J. and Wield, D. (2001), PITA Project (Policy Influences on Technology for Agriculture: Chemicals, Biotechnology and Seeds) Final Report. http://www.technology.open.ac.uk/cts/pita/; http://www.supra.ed.ac.uk/Publications/paper22.pdf.

Tait, J., Chataway, J. and Wield, D. (2004), 'Governance, Policy and Industry Strategies: Agro-biotechnology and Pharmaceuticals', Innogen Working Paper 12; http://www.innogen.ac.uk/ownPubs/Innogen_paper_12.pdf.

Tait, J. and Lyall, C. (2001), *Investigation into ESRC-funded Interdisciplinary Research*, Report to ESRC. http://www.supra.ed.ac.uk/Publications/ESRC_report_Interdisiplinary_research.pdf.

Tait, J. and Williams, R. (1999), 'Policy Approaches to Research and Development: Foresight, Framework and Competitiveness', *Science and Public Policy*, **26**(2), pp.101-112.

Tait, J., Williams, R., Bruce, A. and Lyall, C. (2002), *Interdisciplinary Integration in the Fifth Framework Programme (II-FP5)*, Report to EC (Accompanying Measure SEAC-1999-00034). http://www.supra.ed.ac.uk/Publications/FINAL_REPORT.pdf.

Chapter 2

The Governance of Technology

Perri 6

There is a widespread anxiety among pundits, some international institutions and some NGO activists that "globalisation" has made the governance of technological change even more difficult and perhaps impossible. Against this view, this chapter will argue that the death of governance is, like that of Mark Twain, much exaggerated. Specifically, it will show that there is more governance of technology than there has ever been; there is an extensive menu of ways in which transnational governance is being conducted even where global treaties like that signed at Kyoto fail or are never tried; and there is no zero-sum relationship between domestic and transnational governance.

After setting out some necessary definitions, this chapter examines the ways in which technological risks, opportunities and the decline of industries reliant upon technologies that have become obsolescent are managed, first in the domestic and then in transnational contexts. Within these categories, the available tools for governance are considered, and their relative weight and interdependence explored.

Many commentators detect a general trend toward deregulation, the decline of control and the rise of other instruments of governance, especially of inducement through market power and of influence through standard setting. Others suggest that there is a general drift toward a "precautionary principle" which is being used to justify the resurgence of control methods. But the evidence appears to suggest a much more complex picture. First, such tools are always interdependent. Secondly, it seems that every assertion of a single type of tool or structure sooner or later provokes a counter-reaction in favour of others. The resulting system of governance is best understood as a dynamic disequilibrium system, not as a system tending toward a particular equilibrium unless interfered with. In addition, there are limitations to approaches that stress tools and instruments.

This chapter uses the taxonomy of four basic tools (*control*, *inducement*, *influence* and *coping*) as an organising principle, but not as the fundamental driver of the governance of technology (cf. Bemelmans-Videc et al., 1998; Linder and Peters, 1998; Salamon, 2002; Hood, 1983; Etzioni, 1961). Control involves more or less coercive permission, mandation and prohibition or else direct provision; inducement encompasses the incentives of taxes, subsidies, fees and charges; influence covers the whole range of informational tools from standard setting through to exhortation, from reporting practices to public advice campaigns; finally, coping is the routine process of adaptation, absorbing risk and muddling through.

In tackling the question of amenability to governance, this chapter argues that much depends on what expectations one has for the standard of governability. Expectations for transnational governance tend to be developed on the basis of analogies with particular strands and types of domestic governance, which is itself risky in two ways. First, the domestic is not independent of the transnational. Secondly, there is a danger of arbitrariness in the selection of the particular style or structure of domestic governance to be a benchmark.

"Technology" must be understood as more than just "machinery" or hardware or even artefacts more generally because many of the things that people express their concern about when they worry about technology have more to do with practices and *techniques*, such as farming practices or techniques in gene manipulation. Technology, in this broader sense, means the techniques, together with any artefacts required for the performance of those techniques, by which economic, social, artistic, scientific, military and even political activities are conducted. However, not all techniques present interesting or important challenges for governance. At any given time, the techniques which are the subject either of regulation or agonised abstention from regulation, are those *around which economic, social or political change and innovation is organised*, and in particular, those techniques that present the possibility of *changes in the allocation of resources between groups*.

This chapter examines the following three kinds of governance challenges or problems that technologies can present:

- The use or deployment of technologies can represent or cause *risks* of mortality, morbidity, economic insecurity, loss of social status, disorganisation to certain types of social or economic or political organisation. Those who concentrate upon risks define the challenge for governance as risk reduction or risk management;
- The use or deployment of technologies can represent or bring *opportunities* for economic betterment, improvements in physical or mental health, changes in social organisation considered desirable by some, etc. Those who pay most attention to opportunities see a central challenge for governance as the creating of conditions for investment and growth in industries using these technologies;
- Declining returns to investments in established technologies, whether in their earlier phase they represented risks or opportunities or both, can present problems of economic and social *decline* in organisations, industries, cities or regions, which can bring a variety of social, industrial, managerial, economic and political problems. Those who are focused mainly on these problems see the challenge for governance as that of the cost-effective oversight of the social consequences of shrinkage of capacity, employment, physical plant, investment and reallocation of labour and of resources to other activities and regions.

This last reminds us that it is a mistake to think that the challenges for governance are only those of dealing with new technologies.

Among the key governance challenges raised by technological risks are the demands and pressures for greater transparency and intelligibility of new technologies to wider publics and special interests groups, and for greater accountability of the main commercial or governmental or professional users of technologies. The risks that are of greatest significance for transnational governance are those that have high probabilities of adverse consequences for people living in jurisdictions other than the ones in which the cause of the risks is located.

The management of opportunities associated with the deployment of technologies is not a cost-free process. Likewise, managing decline and market exit of industries using old technologies is complex and multi-faceted. Thus, in each of the three categories of governance problem, the *speed* of technological change is a crucial issue. The management of risk often raises the question of whether new technologies perceived to be risky can be slowed down. By contrast, efforts in regional policy and industrial policy to manage opportunities typically involve efforts to speed up both the research and development and the diffusion phases of technological change. There are sharp disagreements about whether it is more advisable to accelerate or brake the processes of decline in industries using outdated technology.

The tools of governance are quite generic in character, and so are their limitations and interdependencies. There are actually rather few governance problems that are specific to technology. In part, this is because technology is necessarily to be considered as such a broad category encompassing physical artefacts through to complex financial instruments. Therefore, a discussion of the challenges facing the governance of technology is really only an illustration of the challenges for governance in general.

This set of distinctions provides the basis for the organisation of the analysis that follows which deals respectively with domestic and transnational governance of technologies. Within each section, there is a discussion of trends in the use of control, inducement, influence and coping, and within each of these categories of styles of governance, some attention is paid briefly to each of the three substantive topics of the governance of risks, opportunities and decline.

Domestic Governance of Technologies

Control

Methods of control involve the use of substitution for private provision, and regulation including mandating, commanding, prohibiting and permitting certain activities. Almost inevitably, this is the remit of the public sector, at least in reasonably developed states. Commercial organisations can exercise control over others only in situations of non-contestable and politically secure monopoly or oligopoly. However, behind such situations invariably is to be found the action and inaction of states in deciding to prefer to abstain from the aggressive use of anti-trust policy and law, and to promote or protect the situation of monopoly or oligopoly. In the same way, control methods are available to professional institutes and other professional bodies, but only where they have been granted by governments – typically in explicit statute law – the right to exercise such control.

Risks Domestic governance by control of risks presented by technologies has been one of the fields of greatest overt and explicit political conflict in the period since 1960. The environmental movement has, since the early 1960s, given articulation to a quite general principle that action to prevent low probability, high impact risks of all kinds should be a significantly higher priority, and that the principle espoused by traditional hierarchical strategies of focusing greater resources on high probability risks is misguided. Under the label of "the precautionary principle" (O'Riordan and Cameron, 1994), the claim has increasingly attracted some scientists too (see e.g. Harremoës et al., 2002), thereby undermining the once widely held view that there is a straightforward alliance between scientific experts and those in government. Equally, the continued public support for pro-growth political parties and the modest support for green consumption habits serve to illustrate the diversity of views among the publics. It is a matter of debate just how "precautionary" actual trends in regulation have been, and whether there is in fact as big a difference between a generally precautionary European Union and a generally less precautionary United States as is sometimes claimed: the best assessments suggest that the contrast is often overstated (Wiener and Rogers, 2002).

Although in theory, precautionary arguments could be used to support designs for very sharply tapered schemes of fiscal inducement, in practice, they tend to be deployed in support of prohibition and mandation. Indeed, it is clear that, despite the rush to precaution in the regulation, for example, of human cloning in the US, in many areas of the regulation of technological risk, there are many more examples of shifts away from control toward inducement. For example, following the Californian experiment, many countries have introduced schemes for tradable pollution rights. "Taxing bads" is a fiscal strategy that may nowhere represent the principal source of public sector revenue, but nonetheless does attract increasing use. The shift toward inducement is by no means universal, either geographically or by field of technological risk. It is more typically used in attempts to exercise limitation over unwanted but expected side-effects of the normal operation of technologies, rather than to discourage the use of certain technologies altogether. Where the normal operation of a technology is regarded as suspect or wrong in principle, traditional control instruments are preferred to inducement.

Approaches to the governance of technological risks that are rooted in control typically, and perhaps necessarily, involve efforts to predict the future development of new technologies, their future uses and applications, and the ways in which both development and use might affect already recognised risks and might bring new risks. In many countries, these programmes of predictions are organised by governments and institutionalised as technology foresight exercises (Gavigan et al., 2001). In recent years, in recognition of the scale of uncertainty (Sarewitz et al., 2000), the dominant approach in technology foresight has shifted from studies that yield a single prediction toward those which offer multiple scenarios, and even the imperative to attach probabilities to those scenarios seems to be less strong. However, this does not necessarily reflect any weakening of the commitment to using the results for control. Rather, legislators and regulators are trying to use multiple scenario studies precisely in order to exercise control in ways that prediction has come to be seen as ceasing to afford.

Three basic strategies of control using regulation are distinguished in the literature (Hawkins, 1984; Reiss, 1984; Hutter, 1996):

- deterrence: regulatory control is pursued through enforcement in individual cases. The relationships between regulator and regulatee are adversarial, litigious and wider effects are assumed to flow from the exemplary effect of enforcement. Commitment on the part of the regulated to the behaviour sought is supposed to arise from the regulatee's fear of sanctions. This can be seen in some elements of UK utility regulation in the early 1990s and more recently of the rail infrastructure; this has been a common style in the US approach to social care standards regulation (Day and Klein, 1987) and in some periods, it has characterised the US EPA. The deterrence strategy is easiest to sustain and to show success with, when a well-resourced regulator deals with a single regulatee in respect of a single variable. It is most legitimate when wider publics and consumers are hostile to or suspicious of the regulatee. However, when these conditions are absent or erode over time after the introduction of regulatory powers, or when the costs of litigious enforcement start to rise, declining returns may set in. Again, although in theory matters might be otherwise, in practice, those who argue for precaution in the content of regulations will typically demand deterrence-based approaches to regulatory strategy;

- compliance: regulatory control is pursed through the provision of advice, assistance, guidance and support in individual cases. Enforcement is used only when all else fails. The assumption behind it is that commitment should arise from the regulatee's raised aspirations for quality and for reputation. Examples include the British traditions in social care home standards regulation (Day and Klein, 1987); financial services regulation; and OSHA in the US. The approach is easiest to use when regulating a large but defined number of regulatees, where the majority are not believed to be deliberately trying to evade regulatory standards, and where the regulator is dependent on regulatees for information which would be denied to a regulator pursuing deterrence-based strategy;

- culture-building: relies upon generalised information-dissemination, the use of persuasion, making appeals, creating systems of accreditation for commercially available training and supporting voluntary standards processes. Conversely, culture-building regulators make sparing use of enforcement, and offer only limited and rather distant individual case advice. The best documented example of this is data protection regulation (identified in 6, 1998). The strategy is easiest to use when a poorly resourced regulator faces an indefinite number of regulatees, where that regulator has vaguely defined standards to enforce, where costs of observing violations are very high, where intelligence gathering by other bodies is limited, where the perceived risks are of low salience with key wider publics so that reputation effects for well-behaved regulated organisations are modest, and where regulatees' motivations and commitment vary. Where a number of these conditions begin to change, we can expect to see increased expectation by key interest groups that the regulator will begin to switch to compliance or even to deterrence strategies.

As conditions change over time, including the conditions of risk perception among key publics, regulatory bodies will tend to shift strategy. They can also combine strategies, either sequentially in response to their perceptions of the behaviour of each organisation, or differentiated by sub-sector (see e.g. Hutter, 1996, ch.8). This is most likely to make sense either where the same regulator must operate in more than one field and where conditions differ between those fields, or where conditions in their field of regulation are unknown, ambiguous or uncertain, or in some intermediate condition between the sets identified above as most likely to lead to the adoption of each.

Because technologies are so various, and because the political conditions within and between countries vary so greatly, it is misguided to hope to make generalisations about whether there is a trend toward or away from any one regulatory strategy. However, a few important strands can be picked out. Anti-trust law is an important aspect of the regulation of technologies, as the US Justice Department case (launched under the Clinton administration) against Microsoft shows. Shortly after taking power, the Bush administration signalled that it would take a much less tough approach to the case, and some elements of it have in fact been dropped. In the EU, anti-trust policy under Mario Monti has been much debated, and there are differing interpretations as to just what strategy the Commissioner is pursuing. However, one plausible argument is that the Commission is using deterrence-like rhetoric, but its case-by-case practice may be shifting toward compliance. The Republican administration in the US has generally taken a pro-growth stance, and has restrained any tendencies toward deterrence-based regulatory strategy that might have remained within the Environment Protection Agency. By contrast, following the Enron collapse, there has been pressure on the financial services regulators in the US to shift away from compliance toward deterrence. The same pressures have been observed in the UK, but the Financial Services Authority has remained solidly compliance-oriented and to some extent culture-building in its approach, using its enforcement powers sparingly and only in cases of egregious wrong-doing where it was felt that compliance strategies had not been sufficiently effective. Rail regulators in the UK have frequently talked of using deterrence-based strategies, but have found that their need to balance the regulation of safety risk with the encouragement of the take-up of new technologies and the limited leeway in handling such a highly politicised market has meant that, in practice, they have operated in much more compliance-like modes than their public rhetoric would suggest.

Opportunities Although recent decades have seen extensive political conflict over the manner in which control should be exercised over risks arising from technologies, few made a principled argument that control of risks should not or cannot feasibly be attempted. Yet this has been the dominant theme of debate about the control of opportunities. The rallying cry of the sceptics has been that "governments cannot pick winners" – meaning that agencies equipped with the means of exercising control are also ones that have the most limited capabilities for identifying opportunities and for enabling individuals and organisations to seize them. Industrial policy is, however, by no means dead in the West. Nowhere have the grander aspirations of advocates of laissez-faire been realised for wholesale governmental withdrawal from intervention in the governance of technological opportunity.

An important remaining instrument of control deployed at national level for the governance of technological opportunity today is that of patent law. The use of patent law in recent years in the field of agronomic and pharmaceutical products has become extremely controversial, not least because it has raised the question of whether what are being patented are really discoveries rather than inventions. Deep political conflicts are beginning to break out between multinational biotechnology companies and the Western countries in which they are headquartered and communities and states in the developing world who see resources that they have understood and used freely for many years being patented in the West.

Another area in which the use and manipulation of regulation are of growing importance is in encouraging investment by companies bringing what are believed to be "sunrise" technologies. This is the issue of land use planning ("zoning" in American English). At its simplest, more land or more valuable land can be zoned for these uses. More subtly, regulations can be relaxed or selectively waived around change-of-use applications. Related powers can also be used often by way of additional inducement rather than control, such as making infrastructure investments in roads or rail connections, drainage, sewerage, or tax breaks. These strategies have been particularly important in securing the inflow of investment in some high technology industries into the Republic of Ireland during the 1990s. In many countries, however, a debate has been conducted about how far competition between cities or counties – or, in federal countries, states – in the use of land use planning prohibitions, mandates and permissions and ancillary inducements represents a zero-sum game for the cities (and the country) as a whole. On the other side is the argument that the engine of economic dynamism, technological innovation and employment creation in recent years has been the sub-region, or the region around a particular city and it has been argued that the importance of these regions and the positive role that land use planning can play in their development show that there is a case for some use of control methods, alongside other forms of governance.

In this local level of industrial policy, control methods are only ever deployed in the governance of technological opportunity alongside and even subordinate to the use of inducements. Without the inducement of government procurement, or tax breaks, or hidden forms of assistance and advice from governments, the use of regulation for licensing, permitting, forbidding and commanding is of limited effect in stimulating the recognition and take-up of opportunities.

In developed countries, the main use that the private commercial sector can make of control, or legitimate coercion, is to exercise it with the permission of states. The most important example of this remains application to the courts for the enforcement of contracts in the case of alleged breach, and for remedy for torts, or, by way of judicial review, for redress against public bodies. Many of these are important both for the governance of risks and for that of opportunities arising from technologies. For example, litigation for infringement of patent is an important case of applying to borrow legitimate coercion from states for the market governance of technological opportunity, just as suing for damages and injunctions for torts is for the market governance of risks. The burgeoning costs of litigation in recent years have led to a growing demand by the commercial sector for cheaper forms of conflict resolution, including a plethora of novel forms of arbitration.

Decline   The governance of decline in industries and in the technologies upon which they are dependent is a field that continues to present policy-makers with difficult challenges. Sudden shrinkage in demand for labour, in particular, poses the sharpest dilemma, especially where there is a widespread acceptance that the industry or the technology probably does not have a long-term future.

With the decline of nationalisation, the main application for control methods in this regard is the possibility of using regulation to try to prevent investors from making the process of plant closure and employment downsizing more rapid than they might otherwise prefer. Indirect methods of using regulation to slow down the rate of decline include the use of tariffs for protectionism for "national champions". In general, policy preferences for more rapid shrinkage will lead to the adoption of inducement methods instead. However, even where control methods are used, they are almost never used alone, at least in the post-socialist era. Indeed, a recent cross-national comparative study concluded that the use of control methods for the management of decline in the steel industry was itself in decline (Dudley and Richardson, 2001).

The use of regulation to stem shrinkage in demand for labour remains in place in some European countries, in the form of general labour market regulation to control procedures and timetables for redundancy and to require certain forms of support to be offered to those being made redundant. Of course, in those countries that operate laws that do in fact restrict investors' choices in significant ways, their application runs far more widely than the governance of decline in respect of obsolescent technologies, but without doubt this is an important area of their relevance.

Inducement

Governments have a wide range of tools for inducement including grants or subsidies, public purchasing and procurement, loan and credit guarantees, tax relief, waivers from and reductions in fees and charges (Salamon, 2002). An even greater range of inducement tools is available to private commercial organisations, if they have the means to deploy them, although all rest ultimately on the contractual system of purchasing and selling.

Risks The use of inducement by government to manage technologically-based risks has increased significantly in recent years, not least in response to the advocacy of inducement as preferable, more legitimate, more respecting of liberty, and indeed more effective than control methods of regulation by a coalition of business interests, policy entrepreneurs, professional economists and others. An extension of the idea behind "sin taxes" on alcohol and nicotine to suspect technologies, the idea of "taxing bads" has become a critical slogan for many of these advocates, especially in the field of technologies that are deemed environmentally polluting. More recently, schemes of tradable property rights to pollute have been introduced for certain kinds of emissions, often with gradual reductions in the volumes of such rights issued year-on-year, in order to induce companies to develop and deploy less polluting technologies.

Where governments use many kinds of inducements, however, their capacity for legitimate control must lie behind the ability to deploy inducement. In a weak sense, this is true of purchasing and contracting between businesses and even their non-contractual relations, because the presence of the law of contract and of tort provides a basis of control. However, it is also true in a strong sense for the use of inducement in regulatory policy. It is typically considered necessary actively to police compliance with the tax system by hiring dedicated inspectors and investigators. Similarly environmental regulators must use inspectors to police tradable pollution rights schemes to minimise the numbers of cases in which firms pollute without having such rights. Indeed, in schemes designed to reduce pollution over the course of years by reducing the volume of tradable rights, the requirements for inspection and active enforcement are likely to rise over time, because the shrinkage in rights will increase the perverse incentives for firms to cheat unless the costs of investment in clean technologies fall at a faster rate than the volume of pollution rights available falls.

Opportunities Inducement is, of course, the principal instrument by which private commercial markets exercise governance over technological opportunity. Governments too use inducements including tax breaks for investment either in capital goods generally or in specific technologies. While the scale of direct subsidy in industrial policy may be falling in the developed world, both direct and indirect subsidies continue to be used and regional policy remains an important aspect of technology subsidy in many countries. Perhaps the most obvious example of governments subsidising a technology tends to be missed out of discussions of this kind because the technology is not particularly new and because the policy issues tend to be framed in other terms. Yet European and North American subsidies to agriculture are, in fact, massive subsidies for the use, and indirectly the development of, very particular kinds of technologies. For most of the post-war period, the bias was toward the subsidy of intensive, mechanised agriculture, but in recent years, the balance has begun to shift in response to pressure from environmentalists, from specialist producers and to some extent in response to changes in consumer demand, toward support for organic farming.

The NGO community has aspired in recent years to become more sophisticated in attempting to mobilise inducement. Strategies include trying to mobilise consumers to use their own spending power, and also to mobilise "ethical investors" to use both exit and voice to gain leverage over companies but in general, the evidence seems to be – unsurprisingly – that green preferences have been most institutionalised where they are tightly linked to reduction in the costs to consumers. Voluntary organisations can try to engage in some initial research and then advocacy on technological innovations of the kind that they wish to support, in order to make the case that those technologies will also bring conventional financial benefits to investors. This latter strategy is pursued, for example, by the Rocky Mountain Institute around a wide range of green technologies, using the argument that their reduced generation of waste means increased and capturable benefits for companies (e.g. von Weiszäcker et al., 1998).

Decline   Inducement is the principal tool by which commercial proprietary organisations exercise influence over the pace at which industries and technologies deemed obsolescent or obsolete decline and exit from the market. Simple exit and capital flight, causing plummeting stock prices and an investment crunch is the means by which decline accelerates. By contrast, bail-outs and rescues of various kinds are the main instruments by which proprietary organisations can slow down the rate of decline and closure. Which will prevail in any given case is often hard to predict. For governments faced with the social and political consequences of declining industries and their technologies, the central dilemma about the use of such powers for inducement as they possess, is a stark one. One option is to subsidise in the hope of slowing down the process of shrinkage, thus releasing workers onto the job market at a rate at which higher proportions of those made redundant might be absorbed elsewhere. The other is to allow market forces to govern the rate of decline, in the hope that fast shrinkage might, for all the pain and dislocation, at least not prolong the agony, and might prevent people from being locked into an industry for a period during which their chances of securing more marketable skills only diminish as they themselves age. Rarely, if ever, can – nor perhaps should – democratic governments take such decisions on the basis of economic factors alone. A wider range of considerations of the balance of social solidarities always weighs with politicians, even if only for the cynical reason that votes hang on the acceptability of their decisions.

Influence

Persuasion, moral suasion, ethical appeal, rhetoric, efforts to "change cultures", advertising, public information campaigns and the like, are such ubiquitous phenomena that it is hard to believe that such large investment would be made in these activities if they were generally futile, or if control or inducement were always and everywhere available and effective alternatives. Yet it is still not uncommon to read commentators in business, government and academic circles dismissing suasion as a futile and misguided thing. If the sceptical commentators can see the supposed futility of suasion, then it remains to be explained just why governments, businesses and voluntary organisations invest so much in it. Of course, people cannot be persuaded of absolutely anything. The conditions under which much can be hoped for from persuasion are quite tightly limited. Much depends on context: the situation of those to be targeted with persuasive information and their prior biases; the robustness against experience and the relevance of the message to be offered; the balance of other constraints and incentives to behave in ways that might undermine the behaviours desired by the would-be persuaders; and indeed the other persuasive information available to those targeted. However, as we have seen, much the same is true of control and inducement: only in quite specific circumstances can they be expected to have their desired results.

Again, suasion alone is rarely effective, or indeed tried. Typically, when governments engage in campaigns to persuade the public to buy less polluting cars, for example, they back up their suasive efforts with control and inducements. But we have seen that control and inducement too are rarely used in isolation, and are typically interdependent. Moreover, it was noted above that control has a symbolic effect that integrates it with suasive influence.

Persuasive influence is important in the armamentarium of governments for the governance of technology, and it is interdependent with the other tools. What persuasion does is to give definition, meaning, clarity and context to the deployment of the tools of control and inducement.

Prices and regulations are, as economists insist, signalling devices. But without also using "voice", payment, "exit" and coercion are blunt signals indeed (Hirschman, 1970). Prices alone are thin signals. Even information about price trends over time alone does not "thicken" the information provided. In order to develop, for example, intelligently grounded expectations about future prices – and such expectations are the foundation of market-based forms of governance – much richer flows of information are required. Markets can by inducement generate incentives for companies to provide some of this information: investment analysis and business research services are flourishing commercial activities. However, the power of incentives to keep information proprietary means that there is often a need for a wide range of other information-providing and persuading institutions, including those of government. The technologies associated with innovations in near-money provide a good example. While central banks can use interest rates as signals, interest rate changes and trends alone are not very informative about financial risk and the possible effects of the development of complex, high risk derivatives influencing market expectations, especially during booms. Therefore, at various times since the 1980s, central bankers have had to engage in very straightforward public persuasion addressed to investors to explain their views about the balance of risks associated with the insufficiently careful use of such instruments.

Risks   With declining public deference in Western Europe and North America since the 1950s – tracked in the many surveys that detect falls in attitudinal trust in politicians and civil servants – governments have been under considerable pressure around their efforts to exercise persuasive influence regarding the adoption, use and avoidance of particular technologies.

In some cases, governments are expected by particular groups to try to amplify risk perceptions. It is interesting that, by and large, in the use of technologies in health and health-related behaviours, governments in the UK and the US have generally felt the need to respond to calls to make efforts in risk amplification, whereas the same governments have until quite recently been more responsive to calls to support risk attenuation in many areas of environmental and industrial technological risk (Löfstedt and 6, forthcoming). For example, successive governments have sought to reassure the public that claims of risk associated with emissions from nuclear power stations, emissions 'from power lines, waste incinerators and so on, are exaggerated.

However, governments are pressed by many interest groups to issue warnings about a great many technologies, to the point that other groups protest about the development of a "nanny state". On the other hand, the accuracy, completeness, appropriateness and principled basis of government information campaigns around the environmental impacts of many technologies are bitterly contested. Advocates of precaution call for warnings on the basis of low probability, high impact risks, while business groups tend to insist that the reverse principle should guide decisions.

Some information campaigns about technological risks appear to be reasonably successful. For example, many of the campaigns run jointly by national occupational health and safety authorities, trades unions and employers' organisations have resulted in downward trends in accidents and deaths in some industries. However, these campaigns have been backed by control-based instruments of regulation and inducements in the form of administrative sanctions and civil damages for employers that do not provide employees with adequate information on risks.

Opportunities Influence through the use of information for suasion is in some ways most highly developed and institutionalised in the governance of technological opportunity. The basis of standard setting at national, European and global levels is in the definition of objectives that the design and performance of technologies should achieve. Typically, governments in capitalist countries play a limited role in standard setting (Brunsson et al., 2000). Particular regulators – for example, those for occupational health and safety, or those concerned with environmental impacts of technologies (or even data protection in the case of new privacy enhancing technologies) – may make representations, but they cannot usually command or veto standard bodies. Large businesses are very important both as applicants for formal standards and in achieving sufficient market power that they can set de facto standards. However, the credibility of the de jure standard setting bodies continues to depend in some significant part on their reliance upon independent professional judgment exercised by engineers of various sub-disciplines, inventors, designers, and analysts. Ultimately, standards bodies cannot enforce their decisions using control methods: the force of their work is almost entirely rooted in persuasive influence.

The formal processes of negotiation through the prescribed fora of standard setting authorities, however, rests on an extensive and not altogether structured mass of informal contact, information sharing, debate and persuasion. This ranges from very private conversations through to debates in the relevant trade press and on the internet about the comparative merits and demerits of rival proposals for standards. Battles between rival putative standards for dominance may be fought intensely both by conventional marketing and through the formal and the informal politics of standard setting bodies (e.g. Misa, 1992; Bijker, 1992).

Today, among the most important standard setting bodies which have no direct access to control methods, but which wield almost irresistible suasive power, are those which govern financial services. Accountancy standards are set nationally but accredited globally, and since the collapse of Enron and WorldCom, the adequacy of nationally set standards has once again become a major political issue.

Accreditation and rating are central instruments for the governance of markets of every kind, including for those for technologies. The international credit rating agencies such as Moody's and Standard and Poor likewise exercise purely suasive influence over markets and firms, and have limited direct control powers, but their determinations carry enormous authority.

There are, however, some intermediate cases. Civil nuclear power is, for example, within most of the developed world, subject to some treaty obligations to submit to control systems of governance from the International Atomic Energy Authority (IAEA), but many of that agency's capabilities are more matters of suasive standard setting than of regulation. The fact that investment in nuclear power plants takes many years means that, in practice, the IAEA role often becomes one of extended advice. So, in this case, the distinction between suasion and control becomes somewhat blurred. Nevertheless, the authority of the IAEA rests ultimately on its informational credibility based on its ability to synthesise the collective wisdom of the engineering profession, rather than on the limited regulatory powers granted to it.

The persuasive credibility of Moody's in the private sector, or the International Standards Organisation in the non-profit sector or the IAEA in the transnational public sector all stem from the institutional partial separation between the persuader and the interests of those to be persuaded and of those about whom persuasive information is to be produced, combined with sufficient connection with those fields to ensure that the body has an expertise or understanding of the issues, grounded in some profession or discipline. "Independence" is a misleading term for this, for the IAEA has governmental and industry representatives within its board and staff, the ISO must work with many internal stakeholders representing external interests, and Moody's own stock is traded like any other. Nonetheless, the combination of partial institutional separation and expertise is critical to establishing the institutional basis for others to believe that a body seeking to exercise influence over technologies might be trustworthy.

The importance of non-profit voluntary organisations as persuaders on risks and opportunities arising in relation to technologies has grown, first because they combine partial institutional separation from the personal interests of their donors, buttressed by law, and secondly because their expertise is sufficiently great that their arguments can be considered both disinterested and authoritative. Indeed, the principal instrument that voluntary organisations can deploy to contribute to governance is their ability to exercise suasive influence. However, the membership and donor base of most non-profit bodies are typically rooted in communities outside the worlds of business, technology, the professions or government. This means that they must be reactive in their persuasive efforts around technology, and so tend to focus on perceived risks rather than perceived opportunities.

Decline In the governance of declining technologies and associated industries, suasive influence typically plays a secondary role to inducement and control.

Governments too must attempt to use persuasive influence in order to ask companies to slow down the rate of plant closure, or to provide support for retraining, but without the ability of offer inducements in the form of waivers of regulations, subsidies or tax relief, their chances of success are low. On the other hand, these tools are mute without being wrapped in persuasion.

Coping

For three reasons, there is much less to say about coping – the business of reacting to events and finding ways to survive in the short term – than about the three more strategic instruments of governance. One reason is that it is simply ubiquitous. It is what most businesses do most of the time, dignified by the name of "emergent strategy" (Mintzberg and Waters, 1994 [1985]). It is also what governments do most of their time, as MacMillan's fondly remembered phrase had it, in answer to a question about what really drives decision-making – "Events, dear boy, events...". There are few differences to report between the basic structures in their coping behaviour. Equally, there are few differences between the forms of coping deployed in relation to risks, opportunities and decline.

The second reason is that coping is the bluntest and least informative form of signalling. Each of the three active strategies of governance are reasonably rich in information content: sticks, carrots and sermons are all powerfully specific in the messages they send, but coping provides very little information that others can use. Its passivity and limited transformational capacity upon technologies and industries are measured by the barrenness of its informational freight. The third reason that there is little to say about it, is that the conditions of its efficacy are almost impossible to specify, for coping has no standards other than bare survival. If that is achieved then coping behaviour bears no systematic relationship to the more specific outcomes that are possible. Coping is what people do when they feel that they can gain no transformative leverage upon their world, and in turn, coping does not itself transform very much.

Nevertheless, for all its near-unanalysable character, coping is hugely important, precisely because of its ubiquity and because there are so many situations in which organisations feel that they can gain very little strategic leverage over technological risk, opportunity and decline.

Transnational Governance of Technologies: Problems of Control

Thus far, the argument has been concerned primarily with the application of the four basic strategies of governance of technologies at the domestic level, although reference has been made to the importance of the transnational level to show that the domestic process of governance cannot be understood in isolation.

It is misguided to think that there is a general zero-sum relationship between the volume of governance conducted transnationally and the volume conducted domestically, because the total volume of governance effort is not fixed: the volumes of both domestic and transnational governance by government using control, inducement and suasion have expanded in tandem and in response to the same basic forces of globalisation. If there are zero-sum games, they are to be found at the level of particular powers that may be given up at one level and transferred to another. To be sure, these conflicts are bitterly fought over, as illustrated by the protracted debates over the appropriate level for the regulation within Europe of such things as competition, environmental impacts of technologies, occupational health and safety and its relationship with wider labour market regulation, and of course, control of monetary policy and its leverage over financial development and stabilisation.

This section will be principally concerned with the special problems of transnationalisation for control. This is not to deny that there are some special issues about inducement and suasion across countries, especially for national governments but, in general, these problems are small by comparison with the problems that transnationalisation brings for the exercise of governmental control. This section will also be concerned mainly with the transnational governance of risk and, more briefly, opportunity, since it remains the case that most governance of decline is still undertaken domestically.

Risks

The central challenges for transnational governance are those of uncertainty of and conflict between jurisdictions, problems in enforcement across frontiers, free-riding by states from commonly accepted approaches, free-riding by unscrupulous technologists and firms, and uncertainties over the scope of policies and laws (6, 2002).

For governments concerned about regulatory standards in the governance of risks in respect of, for example, health and safety or environmental impacts or data protection, one of the key specific concerns is the possibility that competitive pressures for foreign direct investment will lead to what is commonly called "a race to the bottom" – that is, competitive deregulation, driven by pressure from footloose businesses offering investment only in the least regulated states. In effect, a "race to the bottom" dynamic would reflect the capitulation of control before the power of inducement.

However, it is not at all clear that "race to the bottom" pressures have been equally intense in respect of the regulation of all technological risks. That there has been deregulation in some countries for some risks is certainly not to be denied. However, this cannot be wholly ascribed to international pressures. For example, the deregulatory trend under the Reagan administration seems to have had more to do with pressures from domestic businesses which were already investing outside the US, and which did not, as a whole, seriously threaten disinvestments there; indeed, the US economy during the 1980s was in fact rather less dependent on trade with, and investment from, the rest of the world than were many European economies which did not deregulate nearly so far. Vogel (1995) identified the phenomenon of "races to the top" in which states (in this case states within the US) which have achieved high levels of economic development and which can hope that sunk costs will keep businesses already located there from disinvesting, may increase the strength of their regulation, and thereby set a de facto standard that other states feel they must adopt. This has been dubbed "the California effect". Moreover, in many fields where one might have expected that, if the "race to the bottom" dynamic is very powerful, the degree of international competition ought to have led to some very clear examples, in fact, there has been evidence of re-regulation and tightening of standards. Power (1997) argues that auditing requirements have risen sharply in many financial services industries, although he does also suggest that actual enforcement and efficacy of this regulation may be much more limited than appears at first sight. Moreover, as Golub (2000) argues, global institutions and indeed continental ones such as the EU have provided quite important brakes upon "race to the bottom" dynamics.

All this suggests that, demanding as the four underlying problems presented to control by transnationalisation certainly are in many fields, national governments are not without some means of tackling them. Golub (2000) rightly points to the role of formal multilateral institutions in some cases as providing national states with institutional ropes to break their fall, should they be tempted to fall in regulatory standards. But it is also clear that formal multilateral institutions have their limitations. They are coalitions of the willing, and sometimes not everyone is willing, even within the developed world: the refusal of the US administration under George W. Bush to accept the Kyoto protocol on climate change is clear evidence of this. Moreover, they are very costly to negotiate, often requiring huge conferences with vast prior exercises of diplomacy by armies of "summiteers". This means that states can hardly undertake the use of this approach for every kind of technological risk, still less every significant issue for transnational governance. Finally, the nature of a negotiation process involving states from many different continents with very different living standards, expectations, institutions, interests, traditions, climates of public opinion, etc, is that formal multilateral treaties have to be designed around the lowest common denominator. Only rarely, on a narrow range of economic issues and with great effort over many years, can higher standards be agreed more generally using such formal and explicit methods: the GATT rounds that led to the creation of the World Trade Organization were in some senses an exception, not a generalisable model. Moreover, the power of governance that the WTO has rests less on its ability to enforce directly, using control methods, its determinations in particular disputes, than in its suasive influence and its ability to mobilise opinion and shape reputations, just as the international credit rating agencies do. Clearly, to the extent that there is co-ordination between states and continental groupings such as the EU and NAFTA, then it cannot be sustained, relying exclusively on institutions such as the World Trade Organization.

Figure 2.1 sets out a taxonomy (taken from 6, 2002) of the basic repertoire of tools for tackling the four central challenges of transnationalisation, illustrated with examples from various information risks arising on the development of new global information networks.

Considering the trends in multilateral relations in the transnational co-ordination of control, it would not be at all an adequate account to suggest that fields of regulation can be divided into those which exhibit "race to the bottom" and those which exhibit "race to the top" dynamics. For most, changes in the standard-setting element of regulation, at least in medium-sized developed countries, seem to be driven by domestic forces, not by international ones.

Certainly, the growth in the numbers of international treaties and intergovernmental organisations suggests that formal hierarchical methods continue to be used where agreement can be secured. Within continents, the evidence of the EU, NAFTA, ASEAN, Mercosur and other such bodies of supra-national treaty-based law making all suggest that at this level, nested regulation and mutual recognition for high salience risks, mainly for unwanted but expected side-effects of normal operations of technologies, can be achieved for some risks at levels of transaction costs that citizens have been persuaded to bear.

Moreover, in addition to formal and hierarchical multilateral treaty systems, there has developed in recent years a plethora of more informal, horizontal[1] systems of co-ordination between regulators in different jurisdictions. Just as many commentators tend to dismiss the weaker tools of governance such as suasive influence as either impotent or blunt, so commentators of the same bias will often dismiss these less explicit and formal styles of co-ordination as insufficiently disciplinary or authoritative to be effective.

The sceptics can, however, be answered. The first point is to stress that the supposedly strong tools of multilateral formal treaty and regulation in fact depend just as heavily upon suasive influence, the manipulation and shaping of information to influence reputations, on the day-to-day sharing of informal and unstructured information, as do the supposedly weaker tools of horizontal syndication between jurisdictions. A second stage in the argument is to point out that the lower transaction costs of horizontal and informal co-ordination enable these methods to be deployed much more widely.

---

[1] "Horizontal" structures are contrasted with hierarchical ones because their affinities with underlying solidarities are with those of individualism and/or enclave: see Douglas, 1982a, b, 1992; Thompson et al., 1990.

Type of ordering	Examples	Cases
A.1. Hierarchically harmonised	A.1.a. supranational law- and scheme-making: setting of laws by international authority e.g. international treaty or other formal multilateral adjustment (Wiener, 1999)	WIPO, UNCITRAL, WTO; aspirations for Council of Europe treaty on cyber-crime
A.2. Hierarchically nested – public	A.2.a. nested regulation: harmonisation of regulatory standards through international agreement, enforced nationally, within which national systems operate and with which the details are expected to comply, and where sanctions and incentives are in place (Aggarwal, 1998: for an example from another field, see Vogel, 1998)	OECD statements on digital signatures and encryption, European Convention on Human Rights; within Europe, much of the EU Directive on data protection
A.3. Hierarchically nested – public / private	A.3.a. supervised self-regulation: encouragement of transnational self-regulation: for example, the encouragement of voluntary standards bodies, backed up by the longstop threat of statutory regulation in the event of the failure of self-regulation, as in the case of domain names control, or quasi-voluntary systems of alternative dispute resolution (Perritt, 1997)	ICANN; sections in EU Directive on data protection creating approval mechanism for private codes of practice
B.1. Horizontally geographically separated	B.1.a. catchment area regulation: agreement separately to regulate, either through executive regulatory agencies or in the courts, transnational flows on their domestic manifestations, either on the destination or source principle, but agreeing to eschew attempts to enforce national laws extra-territorially; geographical restrictions on diffusion (e.g. nuclear weapons proliferation)	Data protection, consumer protection

Figure 2.1   Tools for transnational governance of risks by control

Type of ordering	Examples	Cases
	B.1.b. comity, or permitted extra-territoriality in case of recognised greater interest: a right for regulators from one state to enforce in another if they can show the greater interest than that of the state in the jurisdiction of which they seek to enforce (Johnson and Post, 1997); this is not unlike the principle behind extradition in criminal matters	None as yet, but some extra-territorial applications of US and EU law might be come to be stabilised under such a principle
	B.1.c. regulation lending, or agreement by one state to adopt another's regulatory standards specifically in respect of flows of information trade between the two countries	1999-2000 "Safe harbour" agreement between US and EU on US companies conforming to EU data protection law when providing services using European citizens' personal data
B.2. Horizontally syndicated	B.2.a. syndication of regulation: national regulators engaging in mutual adjustment, dialogue, co-ordination on timing of enforcement, sharing information about violators and new technologies to be promoted, taking each other into account, agreeing joint priorities, creating international colleges of regulators to conduct joint training, policy development and advice	International Conference of Privacy Data Protection Commissioners; regulator conferences for telecommunications, and within Schengen system.
B.3. Horizontally competitive	B.3.a. mutual recognition, mutual adjustment and toleration of some jurisdiction shopping: here, within a span of broadly similar national laws, there is some discretion both upon regulators and regulated about where enforcement is done	Some inter-corporate contract and tort litigation

Source: 6, 2002.

Figure 2.1   Continued

However, part of the reply to these sceptics turns on the question of just what are reasonable expectations for the efficacy, disciplinary power and authority of regulation in general. More exactly, it turns on the question of reasonable expectations for trans-jurisdictional regulation in fields in which there are very large numbers of regulated organisations, where regulators are dependent for information on the regulated, and in which there is a mix of independent incentives for compliance and free-riding both by states and by regulated individuals and organisations.

It is wholly misguided to expect complete compliance with any regulatory system even in the domestic context. Typically, transnational regulation must rely more upon persuasion and some infrastructure and limited inducement, than on capacities for direct control. Here, then, a more reasonable set of expectations for regulation would be based on a domestic analogy of data protection or health and safety at work, where the regulator has multiple objectives defined vaguely, limited resources and limited intelligence-gathering capability, low political salience, an indefinite number of organisations to regulate, and can only use coercive powers as a longstop, but must place greatest emphasis on persuasion. In such circumstances, complete enforcement is not to be expected. Rather, such regulators aim to detect and deter the most egregious violations, respond as best they can to (in effect, cope with) issues and events, and seek to influence the wider commercial culture of respect for health and safety, environmental consideration or privacy. Regulators of these kinds cannot expect to ensure that no one violates the principles. They can be expected to make best efforts to work with the media, with major companies and with consumer activist groups, to ensure that the problem of violation is maintained at a manageable level and to prevent outbursts of excessive enthusiasm leading to regulatory action that is self-defeating or that leads to rules becoming dead letters.

It would be misguided, too, to expect wholly to overcome any of the basic problems of jurisdiction, sovereignty, free-riding and relevance. What can be hoped for is that some of these tools – and perhaps horizontal syndication in particular – may mitigate some of the more severe problems, at least within the developed world within which most investment capital flows. Combating free-riding in technological governance by highly challenged states in the developing world depends on first achieving political stabilisation or integration into the wider community of the developed world, their economic improvement and improvement of the life chances of their most aggrieved groups (not necessarily or even typically the poorest), before it can be expected that they will participate in (for example) the syndicated regulation of technological risk. In the short run, the inability of failed states, chronically poor states and those threatened with exit from the developed world to offer sufficient security and rewards to attract scientists and engineers and investors who might be interested in developing controversial or dangerous uses of civilian technologies provides some security.

Opportunities

Transnational governance of technological opportunity using means of control is much better developed than is transnational governance of technological risk by the same means. In large part, this is because of the close imbrication of control with the opening up of markets to allow inducement to carry the greatest burden of the governance of opportunity: this is of course achieved by state action in regulation and de-regulation.

Without doubt, the most important element of the transnational governance of opportunity has been the opening up of national markets to the free movement of capital, labour, services and goods, increasingly governed using control means by the disciplines of the World Trade Organization. Deregulatory and fiscal competition to attract investment by companies using particular technologies is another example. The use of harmonisation of standards through the ISO and other standards bodies and mutual recognition within the EU and elsewhere of alternative technology standards is a further example of the use of a combination of control and influence methods, in order to open up the scope for a market in inducement. On a bilateral basis, the institutions of overseas aid have been important, because interstate aid to developing countries has sometimes involved technology transfer, or the giving of aid with conditionalities about the use of technologies available from the aid-donating country. Within the developed world, there continue to be important examples of bilateral collaborative initiatives in technological development. The Anglo-French collaboration on Concorde in the 1960s is well known, but during the 1990s, the intercontinental collaboration on the human genome project was backed by many governments through subsidy and encouraged through diplomatic channels. Within the military field, there are many cases of the combination of regulation and public procurement to secure international collaboration in technology development, especially in artificial intelligence applications.

The advent of the WTO, with its powers of dispute resolution to enforce the principle of the progressive removal of both tariff and non-tariff barriers to trade, represents unquestionably a major shift in the balance of institutional weight between the governance of technological risk and the governance of technological opportunity. If its treaty basis were on every occasion to be interpreted in a very strict way, then it could be used in ways that would make illegal the use both of control and indeed of many inducement methods for the governance of technological risk. For example, is it compatible with the founding treaty for a country or a supranational trading block such as the European Union to prohibit the commercial production or import of genetically modified foods?

Lawyers will of course disagree on the interpretation of the basic GATT treaty and the subsequent agreements. What is clear is that, like any law, and in particular any law with very general aims of a constitutional nature, the WTO treaties and agreements are not transposed into practice by an automatic process. Rather, the implementation of treaties through the negotiation rounds, and through the dispute resolution and arbitration process, is a complex and highly politically mediated system, in which many side-bargains are struck and defaulted on between major trading and negotiating players, in which many additional agreements are needed. Moreover, the lawyers have room to disagree over interpretation because the treaty and the subsequent agreements have many, quite widely drafted derogations and exceptions, the use of which in negotiations between states or with arbitrators cannot be predicted in advance. What the WTO calls "umbrella clauses" in the GATT permit national or continental regional regulation on various environmental and health grounds; some others permit subsidies in certain types of case. Of course, some environmentalists would argue that these are insufficiently widely drawn while some business leaders will argue that, on the contrary, they are too widely written. In many cases, the standards to which WTO rules appeal are ones set by other global bodies with standard-setting and, in some cases, de facto regulatory powers.

Both the drafting and the actual process of negotiation rounds and of political manoeuvring during disputes provide a structure that forces disputants to reason with one another, and that provides them with institutional support for coming to settlement. In this context, it would be surprising if a dynamic were unleashed for comprehensive deregulation that did not provoke a counter-dynamic to protect that regulation, and which would eventually be articulated through the WTO system. This is not to deny the point already made above that the balance has shifted toward institutional buttressing for the pursuit of opportunity. However, it does suggest that the GATT and the WTO are unlikely to prevent altogether the national, the regional or indeed the transnational governance of technological risk by methods of control.

Conclusions

This chapter has been organised in part around a distinction between domestic and transnational governance, but has acknowledged both that much domestic governance is shot through with transnational influence, and that "transnational" is a category that spans everything from informal bilateral relations between neighbouring states over the management of land or sea borders, through supranational systems at the level of the continental region such as the EU or NAFTA, to global systems including the informal and horizontal and the formal and hierarchical. It has been argued too that there is no general zero-sum relationship between governance at different tiers: more supranationalisation does not automatically mean less national regulation.

The argument has focused on four key instruments of co-ordination – namely, control, inducement, influence and coping (although coping is so ubiquitous and eternal but so limited in its signalling content that there is less to say about it). The argument has considered three broad fields of challenges for governance – namely, the governance of technological risk, opportunity and decline. Figure 2.2 presents some of the main trends that have been detected in the use of these instruments although it has been stressed that all these instruments are interdependent.

Figure 2.2 does serve to highlight some important issues. It is perhaps remarkable that only in one cell (direct control for the governance of technological decline) is it possible even to suggest, albeit tentatively, a general decline in the use of an instrument. This fact, taken together with the argument of interdependence between tools, and the obvious interdependence between the fields of challenge, should suggest that the total volume of activity in the governance of technologies is, if anything, growing as the absolute size of the world economy increases and as technological development expands.

A central part of the present argument is that we shall not understand how the governance of technology works or how it could work differently, if we do not see the whole system. To examine only control, or only inducement, will yield not only an incomplete account, but in all likelihood, an inaccurate one: only when we see the dependencies that control has upon inducement, influence and coping can we make sense of control methods.

Instrument for governance of...	Control	Inducement	Influence
risk	Continuing use, contrary claims of general decline Transnationally, used in limited numbers of areas for multilateral treaties, but with high transaction costs Growing trend toward weaker practices of syndication	Growing use for "taxing bads"; subject to declining marginal returns	Governmental use, if anything, growing in response to pressures, but controversial Still relatively weak as a tool for the public sector in transnational contexts
opportunity	Shifting away from direct prohibition to indirect to local zoning, patent law, anti-trust, levy laws, Main cases of direct use in defence material purchasing and export control Transnationally, hugely important in opening up markets	Critical and ubiquitous in private sector, domestically and transnationally Some decline in public sector "picking winners" Marginal in the NGO sector	Increasingly important in the standard-setting process and in market-making information: e.g. accreditation and rating
decline	Declining importance?	Continuing great importance	Used to give meaning to other tools

Figure 2.2   Summary of trends in the use of key instruments of governance

It has been argued that a major issue in understanding the nature of the governance of technology is the challenge of understanding and explaining trends in the balance between the governance of risk and the governance of opportunity. Again, however, it has also been suggested that the relationship is not necessarily or even typically a zero-sum relationship, either in the distribution of effort in control, or between control and inducement. For example, just because there is a trend toward greater business power and freer markets, it does not follow that there is not also extensive re-regulation being undertaken, or new forms of governance exercised. Understanding better the relationship between the governance of risk, opportunity and decline is thus an important task for understanding the dynamics of governance.

Acknowledgement

This chapter draws on a much longer ESRC-supported conference paper commissioned by the James Martin Institute for Science and Civilisation at the University of Oxford, given in Lisbon in February 2003.

References

6, P. (1998), *The Future of Privacy, Vol I: Private Life and Public Policy*, London: Demos.

6, P. (2002), 'Global Digital Communications and the Prospects for Transnational Regulation', in Held, D. and McGrew, A. (eds), *Governing Globalisation: Power, Authority and Global Governance*, Cambridge: Polity Press.

Aggarwal, V.K., (ed.) (1998), *Institutional Designs for a Complex World: Bargaining, Linkages and Nesting*, New York: Cornell University Press.

Bemelmans-Videc, M.-L., Rist, R.C. and Vedung, E., (eds) (1998), *Carrots, Sticks and Sermons: Policy Instruments and their Evaluation*, New Brunswick: Transaction Books.

Bijker, W.E. (1992), 'The Social Construction of Fluorescent Lighting, or How an Artefact was Invented in its Diffusion Stage', in Bijker, W.E. and Law, J. (eds), *Shaping Technology/Building Society: Studies in Sociotechnical Change*, Cambridge: Massachusetts Institute of Technology Press.

Brunsson, N., Jacobsson, B. and Associates (2000), *A World of Standards*, Oxford: Oxford University Press.

Day, P. and Klein, R.E. (1987), 'The Regulation of Nursing Homes: A Comparative Perspective', *Milbank Quarterly*, 65(3), pp.303-347.

Douglas, M. (1982a) [1978], 'Cultural Bias', in Douglas, M. (ed.), *In the Active Voice*, London: Routledge and Kegan Paul.

Douglas, M. (ed.) (1982b), *Essays in the Sociology of Perception*, London: Routledge and Kegan Paul.

Douglas, M. (1992), *Risk and Blame: Essays in Cultural Theory*, London: Routledge.

Dudley, G. and Richardson, J. (2001), 'Managing Decline: Governing National Steel Production under Economic Adversity' in Bovens, M., Hart, P. and Peters, B.G. (eds), *Success and Failure in Public Governance: A Comparative Analysis*, Cheltenham: Edward Elgar.

Etzioni, A. (1961), *A Comparative Analysis of Complex Organisations: On Power, Involvement and their Correlates*, New York: Free Press.

Gavigan, J., Zappacosta, M., Ducatel, K., Scapolo, D. and di Pietrogiacomo, P. (2001), 'Challenges and Priorities for European Research: A Foresight Review', *Foresight*, 3(4), pp.261-271.

Golub, J. (2000), 'Globalisation, Sovereignty and Policy-making: Insights from European Integration', in Holden, B. (ed.), *Global Democracy: Key Debates*, London: Routledge.

Harremoës, P., Gee, D., MacGarvin, M., Stirling, A., Keys, J., Wynne, B. and Vaz, S.G. (eds) (2002), *The Precautionary Principle in the 20th Century: Late Lessons from Early Warnings*, London: Earthscan.

Hawkins, K. (1984), *Environment and Enforcement: Regulation and the Social Definition of Pollution*, Oxford: Clarendon Press/Oxford University Press.

Hirschman, A.O. (1970), *Exit, Voice and Loyalty: Responses to Decline in Firms, Organisations and States*, Cambridge: Harvard University Press.

Hood, C.C. (1983), *The Tools of Government*, Basingstoke: Macmillan.

Hutter, B. (1996), *Compliance: Regulation and Environment*, Oxford: Clarendon Press/Oxford University Press.

Johnson, D.R. and Post, D.G. (1997), 'The Rise of Law on the Global Network', in Kahin, B. and Nesson, C. (eds), *Borders in Cyberspace: Information Policy and the Global Information Infrastructure*, Cambridge: Massachusetts Institute of Technology Press.

Linder, S.H. and Peters, B.G. (1998), 'The Study of Policy Instruments: Four Schools of Thought', in Peters, B.G. and van Nispen, F.K.M (eds), *Public Policy Instruments: Evaluating the Tools of Public Administration*, Cheltenham: Edward Elgar.

Löfstedt, R.E. and 6, P. (forthcoming), 'Bring Down the Berlin Wall! Fragmenting and Integrating Environmental and Technological Risk Communication with Health Risk Research', submitted.

Mintzberg, H. and Waters, J.A. (1994) [1985], 'Of Strategies, Deliberate and Emergent', *Strategic Management Journal*, 6, pp.257-272, repr. in Tsoukas, H. (ed.) (1994), *New Thinking in Organisational Behaviour: From Social Engineering to Reflective Action*, Oxford: Butterworth-Heinemann.

Misa, T.J. (1992), 'Controversy and Closure in Technological Change: Constructing "Steel"', in Bijker, W.E. and Law, J. (eds), *Shaping Technology/Building Society: Studies in Sociotechnical Change*, Cambridge: Massachusetts Institute of Technology Press.

O'Riordan, T. and Cameron, J. (eds) (1994), *Interpreting the Precautionary Principle*, London: Earthscan.

Perritt, H.H. Jnr. (1997), 'Jurisdiction in Cyberspace: The Role of Intermediaries', in Kahin, B. and Nesson, C. (eds), *Borders in Cyberspace: Information Policy and the Global Information Infrastructure*, Cambridge: Massachusetts Institute of Technology Press.

Power, M. (1997), *The Audit Society: Rituals of Verification*, Oxford University Press, Oxford.

Reiss A, 1984, 'Selecting Strategies of Control over Organisational Life', in Hawkins, K. and Thomas, J.M. (eds), *Enforcing Regulation*, Boston: Kluwer-Nijhoff.

Salamon, L. (ed.) (2002), *The Tools of Government: A Guide to New Governance*, New York: Oxford University Press.

Sarewitz, D., Pielke, R.A. Jnr. and Byerly, R. Jnr. (eds) (2000), *Prediction: Science, Decision Making and the Future of Nature*, Washington DC: Island Press.

Thompson, M., Ellis, R.J. and Wildavsky, A. (1990), *Cultural Theory*, Boulder: Westview Press.

Vogel, D. (1995), *Trading Up: Consumer and Environmental Regulation in a Global Economy*, Cambridge: Harvard University Press.

Vogel, D. (1998), 'The Globalisation of Pharmaceutical Regulation', *Governance*, **11**(1), pp.1-22.

von Weiszäcker, E., Lovins, A.B. and Lovins, L.H. (1998), *Factor Four: Doubling Wealth, Halving Resource Use*, London: Earthscan.

Wiener, J. (1999), *Globalisation and the Harmonisation of Law*, London: Pinter.

Wiener, J.B. and Rogers, M.D. (2002), 'Comparing Precaution in the United States and Europe', *Journal of Risk Research*, **5**,(4), pp.317-350.

Chapter 3

The Governance Challenges of Breakthrough Science and Technology

Graham Spinardi and Robin Williams[1]

Steering the Next Technological Revolution with Hindsight from the Last? An Introduction to the Problem

Science and technology policy (STP) represents a somewhat difficult area for policy-makers. It deals with arcane domains which are not readily understood by policy-makers or wider publics, who largely lack the requisite backgrounds to assess technical claims. At the same time, science and technology (S&T) are seen as important for society and the economy. Thus, we find a recurrent fear expressed across many STP statements about the risks of nations and regions getting left behind in the global technology race. Alongside this, some new and emerging technologies have been the focus of sharp concerns about unanticipated and unwanted consequences for health, the environment and for social and ethical values. The former arguments have been used to legitimate increased public sector research investments. But this raises questions about how these funds might be allocated – and how the effectiveness of these investments can be audited and assessed. Are these investments delivering value for money – and are they delivering the best value for money? The difficulty is that the outcomes are not known with any certainty in advance. Science is serendipitous. Some research will have only modest results, other work will lead to breakthroughs in scientific understanding; and sometimes the innovations that result will have widespread applications with far-reaching economic and social implications (solid-state physics and the microprocessor revolution as a case in point (Braun and MacDonald, 1982)).

[1] This chapter draws in part upon discussions in the AT-BEST Project 'Assessment Tools for Breakthrough and Emerging Science and Technology' European Union Sixth Framework Programme, New and Emerging Science and Technology (NEST-2004-508929). The project was led by Prof Arie Rip at the University of Twente, and also involved Prof Joyce Tait (Edinburgh), Philippe Laredo and Aurelie Delemarle (ENPC, Laboratoire Techniques, Territoires et Sociétés) and Tilo Propp (Twente). We would like to acknowledge the important input of ideas from our AT-BEST colleagues (particularly Arie and Joyce). However, responsibility for any failings rests with the authors. The chapter does not represent the views of the European Commission or NEST.

One answer to the non-transparency of technical fields might be to get scientists and engineers to carry out resource allocation – and peer review has played a major part in selection procedures for research funding. Such a move is not without difficulties: delegating decision-making may tend to favour established disciplines and fields of study over emerging ones (and indeed does not resolve the problem of dividing money between fields). There have been long-standing concerns that it may equally prioritise scientific advance over its exploitation. There are other interests (commercial, state and public) that may also need to be addressed. On the other hand, civil servants are not well-placed to make these kinds of decisions. Indeed in areas of technology policy, the failure of a number of high-profile, high-technology champion projects has led (particularly in the UK) to a view that the state should not get into the risky game of "backing winners". And today we find the increasing dynamism of science and technology involves not just an accelerating pace of new knowledge generation, but also increasing turbulence, with some areas slow to deliver and others yielding new understanding and new applications faster than any had anticipated. Policy-makers have been forced to rethink traditional concepts of planning and control of developments in favour of more flexible and opportunistic approaches – a change which Brosveet and Sørensen (2000) have characterised as "Fishing not farming".[2]

The implausibility of micro-management of scientific and technological developments may not be problematic if the broad policy frameworks are in place to motivate the generation of new knowledge and its commercial and socially beneficial exploitation. The key issue, which this chapter focuses upon, is that we are forced to address the next generation of science and technology armed with the strategies, tactics and battle-plans that were successful in the last revolution. Policy frameworks are based upon paradigms developed, tested and refined in earlier rounds of scientific and technological advance. They may still be effective where further developments broadly fit within past patterns and parameters. However, new and emerging science and technology may equally require rather different support mechanisms, different evaluation criteria and different tools for intervention. This is the challenge posed by discontinuity in science and technology. Discontinuities do not just result from changes in the objects of study and specialist knowledges deployed thereon. Developments in science and technology may also involve new actors and different relationships between actors (for example, changes in the range of disciplines deployed and in linkages with industrial players); different ways of organising science and technology (for example, increases in scale of scientific research and/or of technology exploitation activities); and different broader regimes (for example, of risk regulation and intellectual property rights).

This chapter reviews the policy challenges posed by such discontinuities and, in particular, by so-called "breakthrough" science and technology. It notes the

---

[2] Whereas "farming" describes 1980s national strategies involving the planned development and growth of technological capabilities in core technological fields, "fishing" implies a selective strategy, drawing on offerings and opportunities found in local and global technology markets.

different dilemmas that arise (for example, between alignment of effort and the maintenance of diversity) and discusses some of the frameworks that have been advanced to tackle these challenges. Finally, it considers some of the methods of assessment and decision tools and criteria that may be needed for breakthrough science and technology.

Discontinuity in Scientific and Technological Development

Discontinuity in the paths of scientific and technological development has long been recognised as presenting a difficult challenge to public policy for science and technology (and equally to business strategies and public responses). Although much science and technology can be characterised as "normal", where change involves incremental improvement based on established "paradigms", and is therefore reasonably predictable and relatively easy to manage, "radical" breakthroughs in science or technology do not build incrementally on existing paradigms and are thus difficult to predict and to manage (Freeman, 1974).

However, the importance of paradigm shifts has long been recognised. New industries are formed and old ones displaced by what Schumpeter famously described as "creative destruction". Paradigm shifts occur due to innovations that "command a decisive cost or quality advantage and that strike not at the margins of the profits and the outputs of the existing firms, but at their foundations and their very lives" (Schumpeter, 1942). A cyclical model of technological change encapsulates this process (Anderson and Tushman, 1990). Emerging S&T can lead to a period of what Anderson and Tushman term "technological discontinuity" during which previous dominant approaches are replaced by a new paradigm. According to this cyclical model, "sustaining" technologies are those that enable the continued incremental improvement of an existing paradigm, whereas "disruptive" technologies are those that lead to the establishment of a new paradigm.

Breakthrough and emerging S&T thus pose a major challenge for industry. Firms must pursue continuous improvement to make the most out of an existing paradigm, while at the same time searching out new opportunities that may accrue from emerging S&T. Policy-making also needs to be attuned to the significance of paradigm shifts in S&T as these can call into question existing criteria for assessing technology, and expectations about its social and economic outcomes and implications. For example, it has been suggested that new applications of genomics call for rather different support mechanisms (notably involving very much greater levels of financial support) than is provided by the UK's current mechanisms to support biotechnology commercialisation. [3]

---

[3] For example, the Director of the newly formed Bioinformatics Centre at the University of Edinburgh, Peter Ghazal, argues [personal communication, 2001] that the level of resource needed to achieve and to spin-out innovations in life sciences had increased by at least one and perhaps two orders of magnitude with the shift from Biotechnology to "Genomics".

This challenge may be exacerbated within the context of an emerging knowledge economy – with a greater volume of scientific and technological research and its more rapid communication. For example, we see across a number of fields, advances in scientific understanding and techniques that would appear to have potential for widespread application in medicine and many other areas of human activity with the promise of economic opportunities and potentially far-reaching implications for the quality of life. Greater dynamism in the knowledge system seems to be one of the factors stimulating disciplinary "churning" with increasing levels of interaction between disciplines characterised by the increasingly rapid formation of sub-disciplines, notably in and around the life sciences and nanotechnologies, with new fields of enquiry emerging at the interface of existing specialisms (for example, bio-informatics, proteomics), as well as shortened knowledge life-cycles as attention moves away from established domains. There has been, at the same time, growing recognition of the need for concertation of effort in relation to both achieving advances in knowledge and in pursuing its societal application and commercial exploitation. Thus, we see attempts to build "platforms" for research geared towards both increasing the scale of research activities and the complementarity of capabilities (though these are conceived differently, for example, for knowledge advance and exploitation). This is associated with the development of integrating STP frameworks; for example, current conceptualisations of research priorities around a limited number of poles such as Nanotechnology, Biotechnology, Informatics, Cognitive Science (NBIC – also summarised as Atoms, Genes, Bits, Neurons) looking towards some kinds of convergence between fields (National Science Foundation, 2002).

The dynamism of S&T knowledge and the turbulence of fields of enquiry pose a problem for conventional science and technology policy. The dominant contemporary approach has sought to go beyond earlier dichotomies between "science push" and "market pull" as motors for innovation and has highlighted instead the improved efficiency in innovation that may result from "coupling" between S&T producers and users (Freeman, 1974). A key challenge for research policy thus revolves around the assessment of the potential "coupling" between a new approach and a user requirement (either existing or to be created). However, a growing body of research from Science and Technology Studies shows that new S&T advances often do not find immediate application, and sometimes only achieve success in an application quite different from that originally conceived. The innovation pathways and socio-economic outcomes of new technologies are often far removed from initial presumptions. Not only may the eventual benefits differ significantly from the initial promise of a new technological field, but there may also be undesired consequences (and in recent years a combination of factors has heightened fears about unrecognised risks that may arise with new technologies). These uncertainties pose enormous difficulties for science and technology policy, which is charged with the task of supporting research that will bring profound advances in understanding as well as with promoting the exploitation of new S&T knowledge. This is the dilemma at the heart of assessing the potential of breakthrough S&T.

What is "Breakthrough" Science and Technology?

Identifying Paradigm Shifts in Science and Technology

Much of science and technology can be characterised according to Kuhn's (1962) notion of "normal" where change involves incremental improvement based on established "paradigms" and is therefore reasonably predictable and relatively easy to manage. However, "radical" breakthroughs in science or technology do not build incrementally on existing paradigms and are thus difficult to predict and to manage.

The management of innovation literature, to a large extent, has similarly stressed the importance of incremental product development which can proceed within existing routines (for example, Prahalad and Hamel, 1990). However, simply relying on incremental improvement of existing paradigms is likely to provide short-term success only, in the face, for example, of competition from lower-cost economies. However, the potential for breakthrough developments may be overlooked, as companies' routines for selecting innovations may not be applicable to breakthrough S&T, which will often not fit in with existing business operations – for example, typical payback economic assessment is not likely to be useful (O'Connor and Rice, 2001). Today, the distinction is increasingly made between "sustaining" and "disruptive" technologies, with the former enabling the continuous improvement of existing product lines and the latter requiring/causing a major shift in the product and its place in the market (for example, Kassicieh et al., 2002).

Whilst breakthrough innovation within the management literature is mainly theorised at the level of the individual firm, breakthrough technologies (like related concepts of radical and systemic innovations) frequently require collaboration amongst a range of research and commercial players bringing together complementary knowledge and other resources required. So the building of breakthrough technologies may involve building support across a network to sustain and carry forwards the innovation (O'Connor and Rice, 2001), which is likely to involve developments in the core technology and in a range of ancillary innovations needed to get the new system to be effective (Freeman, 1974). Questions therefore arise about how support can be mobilised across such networks – through, for example, the alignment of key firms in a supply chain, or support from influential potential customers, and frequently through government funding and support for knowledge exchange and collaboration which can be critical in providing external validation of a new concept.

In some cases, breakthroughs may emerge as part of a programme of "normal", incremental research, and their significance may only be recognised in retrospect. In many cases, the initial development is not in isolation significant, but rather it is its application across a range of previously unconnected fields that can be seen as constituting breakthroughs. It is not the new scientific knowledge or artefact that is in itself significant, but rather the building of social networks. In other cases, the breakthrough will have been deliberately targeted with the goals set by what are perceived to be existing "reverse salients" (Hughes, 1987). The exact means by which the reverse salient is overcome, and the way in which the challenge is conceived, may not be predictable, but the overall thrust of the research will have a clear agenda.

The transition from one paradigm to another may not always be clear or rapid. Established technology paradigms will tend to have achieved optimised performance from earlier investments (Gilfillan, 1970) and "learning-by-doing" (Arrow, 1962), they will have a mature assemblage of complementary technologies and techniques (Rosenberg, 1976), and they may also have gained strong inertia from "path-dependence" effects (David, 1985). Emerging S&T may offer potential advantages over existing "dominant designs"(Anderson and Tushman, 1990), but it can take considerable further development to realise fully these advantages. The very potential of emerging S&T to establish a new paradigm can make it "disruptive", as opposed to "sustaining" of existing social arrangements, particularly as regards to an established customer base (Walsh et al., 2002).

For these reasons, breakthroughs can often only be seen clearly in retrospect. Identifying significant breakthrough research in real time is more difficult. Scientists and technologists typically structure their work around their own visions of the future (or link their work into broader visions emerging from research policy-makers and funders), putting forward what they see as a plausible projection – that we will refer to as an "expectation trajectory" – for the advancement of their research and its eventual application and use.

Mobilising Promises: Riding the Expectation Trajectory

This articulation of the scientific and technological outcomes and scenarios for its further exploitation is, in part, an intellectual device used to organise work plans, but more significantly it is a necessary part of applying for, and gaining financial support. The mobilisation of expectations – of the promise of technology (van Lente and Rip, 1988) – becomes part of the "currency" of discussions between researchers and funding bodies and also, in turn, between those funding bodies and governmental paymasters who determine the level and shape of state research funding. Technology visions and promises are, of course, boundary objects that signify different things to different players. For the researcher, promise is about securing funding for this project/researcher in competition with another on the basis of its expected outcomes; for the research funder these consolidated promises are mobilised to give assurance to government funding departments that the research programmes will deliver longer-term economic and social benefits and thus can be justified in terms of return of investment in competition with other kinds of demand for public funding.

Obtaining funding for research is a political activity. Researchers must take on the role of "heterogeneous engineers" (Law, 1988) in addressing simultaneously both social and technical challenges. Researchers and, in particular, successful programme managers are attuned to shifting policy goals, funding rules, and assessment methods. While such policy changes can shape the content of research, they may equally have their most immediate impact on the *presentation* of research applications. Thus, the shift in emphasis from basic to applied research in recent years has seen researchers become more adept at pointing up the applications of their work while allowing them to carry on along similar lines as before. Likewise, the huge public focus, and resulting surge in funding, given to such areas as biotechnology and nanotechnology, has resulted in researchers reclassifying their existing work to fall within these suddenly lucrative disciplines.

In the case of academic research, tensions can often exist between the various goals of the researchers and those of the funding agencies. Public funding of research will most likely be justified as important for the innovation necessary for "wealth creation", or perhaps also to provide a public benefit in the form, say, of environmentally cleaner technology. Academic researchers will be aware of these expected outcomes, may be strongly supportive in trying to achieve them, but nevertheless will most often have their primary focus on the practices that govern their career advancement.[4]

Radical developments in S&T, by definition, step outside of existing paradigms, and so require the establishment of a new "expectation trajectory". As van Lente and Rip (1988) describe in the case of membrane technology, expectations of emerging S&T act as a "prospective structure" linking together actors in an overlapping vision (though not one that requires a shared view) and creating nodes of interaction. One way of articulating this process is to see it as a sequence in which first promises are made, leading then to the establishment of agendas, which in turn result in the setting of requirements. This promise-agenda-requirement sequence is at the heart of emerging S&T.

The importance of expectations can be clearly seen within the currently "hot" areas of NBIC, and particularly with biotechnology and nanotechnology. These areas have attracted huge levels of funding, and established large research programmes, based on expectation trajectories put forward by their proponents. Although few successful products exist, the generalised expectation trajectories have been constructed so skilfully as to gain widespread support. But how are these generalised expectations put together around certain fields of enquiry, and what is their relationship with the more specific sets of expectations advanced (and due to be delivered) in the funding of particular projects and initiatives?

There are obvious dilemmas in terms of how to pitch expectations: aim too low and there may be difficulties in justifying public support. However, if claims are

---

[4] In the UK, for example, the Research Assessment Exercise, with its emphasis on the quality of research outcomes (measured primarily in terms of publications in high-ranked journals), may well come into conflict with a more commercially-orientated approach (patent first, publish later). One traditional manner in which the tensions between the goals of scientific excellence and of exploitability has been addressed is through separation of research funding into basic and applied research. With basic research, sometimes described as "curiosity driven" research, funding is often distributed at the behest of the scientific community itself on the basis of quality and reputation, to ensure that key fundamental advances in scientific understanding are not overlooked in the search for the more effective exploitation of knowledge in commercial innovation.

pitched very high there is a danger of a backlash if progress in meeting these expectations is too slow (as happened in the case of artificial intelligence, where funding in the UK and USA was cut back in the 1970s).[5]

Numerous studies from the history and sociology of technology point to the protracted processes of the development and commercialisation, marked by frequent set-backs, of radical new technologies seen with hindsight as revolutionary. Actual experience is out of kilter with the dominant rhetorics of technology supply, however, and proponents of new scientific and technological programmes typically seek to reassure by projecting a "linear" and steady exploitation pathway. There is an obvious dilemma, which proponents of new S&T (from research and policy contexts) must grapple with. The temptations to overstate capabilities and understate challenges may lead to a kind of foreshortened, "compressed foresight" in which the technical and societal outcomes of proposed breakthroughs are projected as arriving in too rapid and straightforward a manner. This carries obvious risks of disillusion. Equally we find that these overstated promises can mobilise anxieties as well as hopes – see for instance the recent debate about nanotechnology and "grey goo" – around the prospects (which in the immediate term seem rather implausible) of self-replicating nano-scale machines (Phoenix and Drexler, 2004). Similar kinds of hyperbolic extrapolation are visible in many of the debates about stem cells and new reproductive biology, focusing for example on the risks of human cloning.[6] Underpinning both the utopian and dystopian futures is a very linear sense of the unfolding trajectories of technology development and of their social, economic and environmental implications.

The specific outcomes of these large programmes of research cannot be reliably predicted, but it can be expected that they would encompass a wide range of activities, not only advancing scientific knowledge, but also producing new instrumentation and techniques. The benefits of such large programmes lie with the scope they afford for the concertation of effort (in relation to both scientific advance and its exploitation). The corollary of this kind of concertation is an inevitable focusing of effort and attention. Such programmes, and the expectations

---

[5]   Thus, for example in the early 1970s, funding for artificial intelligence research in the UK and US was significantly reduced when a UK government sponsored report, known as the Lighthill Report after its main author, suggested that there was little prospect of the rather overstated claims articulated by early AI proponents in the late 1950s being fulfilled (Fleck, 1982; National Research Council, 1999).

[6]   What seems remarkable here is the way in which the lens of "ethical issues" is used to project apparently clear sets of social and moral consequences from the existence and thus imputed application of new biological techniques and knowledge, even though their likely outcomes cannot readily be anticipated. Given the centrality that ethical dimensions are acquiring in many of these debates, a stronger input from empirical studies in S&T development would seem to be called for, that can pay attention to the way in which the technical and societal outcomes of a research programme are not "built-in" to the initial stages of design, but are patterned over time through the multiple choices as different stages in the "biography" of the innovation.

on which they are based, are not open-ended. Particular approaches and conceptualisations will tend to gain preference within a local research team or institute, or a wider funding agency or a political system. Other approaches will thus become neglected. Questions arise regarding how this inevitable focusing of effort and attention on particular pathways can be achieved in ways that do not unduly reduce the space for innovation.[7]

This resort to platforms and large programmes represents, for many fields of S&T endeavour, a new mode of scientific organisation, in terms of the scale of activities and of the level and types of interaction, although these have emerged from recognition that existing models for support of research excellence and its exploitation could no longer be taken for granted. However, the efficacy of these various large-scale funding models, and their appropriateness to particular technological and societal settings, remains essentially unproven. They would appear to offer important benefits in terms of economies of scope and scale and of enhanced learning opportunities within centres of excellence thereby established. There might also be disbenefits for example in a narrowing of diversity and in a move from "competitive" to "oligopolistic" knowledge production. These emerging platforms/programmes, for example in nanotechnology, have been put together in rather different ways, for example, between the UK, France, the Netherlands and the USA depending on their national research policy traditions and mechanisms, and it would be instructive to compare their differential effectiveness.

Given the complexity of the goals surrounding research policy and the lack of an evidence base to validate one framework over others, a key consideration may be to maintain some level of diversity in terms of mechanisms, incentives and frameworks for supporting S&T advance and its exploitation and in terms of the programmes of research and projects supported – what Rammert et al. (2003) have described as *innoversity*. A balance between concertation and diversity is thus required – with the latter being particularly relevant where uncertainties are high regarding the most appropriate pathways to be pursued.[8]

Although not irreversible in principle, decisions to invest in a new field of S&T can in practice create such momentum behind the favoured approach that other

---

[7] Many research programmes start off with rather ambitious generic goals – but as they move to completion there may be a shift towards more concrete outcomes that can be used to justify the investments made. Often there is rather thin evidence of innovative outcomes (e.g. the Alvey Programme – see Oakley, 1990). The question arises whether the search for concrete deliverables that correspond to policy-makers' expectations that research investment will lead to innovation and commercial exploitation will be pursued in a way that might inhibit the actual achievement of innovation in unanticipated ways.

[8] An interesting illustration is provided by the European Union 6th Framework Programme, New and Emerging Science and Technology (NEST) activity (see http://www.cordis.lu/nest), set up to support unconventional and visionary research with the potential to open new S&T fields, in a context of concern that the greater targeting and application focus of the bulk of FP6 research funding might encourage mainstream over promising but untried possibilities.

options are closed off, at least in the short term. This momentum is, of course, a socio-technical construct – there is no inevitability about "progress" in S&T and in the particular pathways followed. Politics, national economic and military interests, industrial competition as well as learning from (and mimicry of) international competitors, all play a role in shaping how such new fields of S&T are created. They do not arise simply from laboratories and the minds of scientists. Nevertheless, the prospective structure created by expectations in an emerging area of S&T leads to "protected spaces", at least temporarily, in which attempts can be made to satisfy the expectations. Governments create programmes, universities set up research centres and industry funds projects, all based around a conflation of developments in S&T into a new field.

New Paradigms for Science and Technology?

These developments across a number of STP frameworks suggest the emergence of new broad frameworks and paradigms for thinking about policy, in particular for breakthrough S&T.

Mode Two S&T

Gibbons et al. (1994) and Nowotny et al. (2001) have suggested a shift in the practice of science away from the traditional paradigm of university and discipline-based Mode 1 knowledge production towards a multi-site, inter- and trans-disciplinary Mode 2 way of working. They emphasise the role of collaboration between different disciplines and between academic and commercial/policy players in relation to both scientific advance and its technological exploitation. This simple schema has been criticised. For example, their analysis fails to capture fully the nature of current developments in scientific research in fields such as genomics, which constitutes, not a shift to Mode 2 knowledge, but rather a return to an earlier approach, that of "natural history", where, the emphasis is not on "classic" hypothesis-experiment type science, but rather on description and classification of genome structures and properties (Rip, 2002). Nevertheless, there is evidence of the break-up of the traditional emphasis on academic disciplines around which domains of knowledge are taught and validated. Mode 2 knowledge spans traditional disciplinary boundaries – highlighting the need for practitioners always to be trading ideas in a domain outside their own specialism. Their view also calls into question the distinctions established in much public policy between basic (or curiosity driven) science, applied (or problem driven) science, and technology. The production and exploitation of knowledge are now interpenetrating. It is notable that the basic/applied research dichotomy is becoming broken down in the new

NBIC STP environment – in which funding for scientific advance and for exploitation are both being targeted towards centres of scientific excellence.[9]

NBIC

Over the last two decades or more, we can see that support for technological innovation has been increasingly generic and programmatic – developments that can be traced back, in particular, to the Japanese 5th Generation computing programme, and the various responses to this by Western governments (such as the "Alvey" programme in the UK (Oakley, 1990)), which involved research and exploitation activities across a range of related fields of information technology.

Today, it is becoming increasingly popular to refer to a "second industrial revolution" based on what is known alternatively as NBIC (nano-bio-info-cogno) or the Little BANG theory (bits, atoms, neurons, and genes). Rather than adapting existing materials and techniques, NBIC is predicted to enable new technologies to be shaped from scratch. Such developments are conceived as being inherently generic in nature as the "convergence of diverse technologies is based on material unity at the nanoscale and on technology integration from that scale" (National Science Foundation, 2002).

This shift in thinking about S&T has two major implications. First, whether or not NBIC is a short-term reality in terms of actual products, it is important in the way that it shapes expectations of future technological trajectories. Second, at the heart of the NBIC is a conceptualisation of S&T as involving a matrix of generic building blocks from which the key to exploitation is the cross-cutting linkages. The NBIC formulation and the development of more programmatic funding (for example, for nanotechnologies) have become a feature of STP across a number of countries – notably the USA, Europe and beyond.[10] These kinds of overarching policy frameworks can help policy-makers and their clients make sense of a turbulent technical setting. However, we must remember that their appropriateness remains unproven. Such common frameworks may help reduce uncertainty, but at the same time they may also constrain thinking and limit policy analysis in unhelpful ways.

This compelling vision of NBIC further carries the suggestion of a breaking down of barriers both between different disciplines and between different parts of the innovation system. We have already noted the increasing "disciplinary churning" in the formation and reformation of new sub-disciplines at the interface of earlier disciplines and fields (though it is by no means certain that current

---

[9] The dilemmas underpinning the basic/applied research distinction resurface, in a different form, in discussions about how to advance nanotechnology (e.g. top-down versus bottom-up nano; and in competition between research and exploitation centred initiatives).

[10] We see here processes of mimicry in technology policy between Europe, the USA and Japan that have been a feature of earlier anticipated technological revolutions (dating back to the Japanese 5th Generation computing programme in the 1980s, and the information infrastructure policies of the 1990s).

developments are indeed a break with the "normal" process of evolution of research domains and sub-disciplines and that the boundaries between basic S&T disciplines are indeed being profoundly reshaped). However, for application and exploitation these research activities must also link in with potential intermediate and final users of this knowledge. S&T must emerge into application domains and settings (what evolutionary economists have termed "the selection environment") – which are likely to remain distinctive in their characteristics. Although emerging S&T may only have weak coupling with potential end-uses, this "selection environment" will nevertheless play a significant role in shaping expectations.

We may compare, for example, the fields of nanotechnology and biotechnology, which differ greatly in the ways that both basic research and innovation are organised.

For example, the current paradigm of pharmaceutical innovation is strongly shaped by regulatory structures and the centrality of patent protection for profitability. These pressures have led to an innovation process in pharmaceuticals which is strongly research driven, and which to a large extent conforms to the linear model, with its implication of a steady and broadly one-way flow of ideas from "upstream" public and private sector research laboratories to "downstream" commercialisation and use, as new chemical entities are tested in order to discover useful therapeutic properties along with, hopefully, lack of toxicity or serious side-effects. Subsequent development of potential drugs follows a fairly predictable trajectory involving a series of animal and human trials, under well-established acceptability criteria, as mandated by the regulatory authorities (Tait and Williams, 1999).[11]

Genomics offers a new approach to drug discovery, based on therapeutic solutions tailored towards particular genetic make-ups, but it is likely that the innovation system, and in particular the system for regulatory approval, will remain broadly the same as conventional pharmaceuticals, with a rather clear innovation hierarchy and broadly predictable regulatory process (though there may need to be some adjustment to the regulatory framework for drugs targeted towards particular genotypes). On the other hand, we can point to the enormous and crucial role of public sector research in advancing core knowledge and as a source of commercialisable applications, driven by expectations of huge economic and health benefits that may accrue.

By contrast, although nanotechnology is being espoused as applicable across an extremely wide range of applications, much of the current research is located in (various) traditional disciplines such as chemistry, physics, electronics and materials science. At this stage there is not an established nanotechnology innovation system – and developments in these areas are likely to have implications for a wide range of industries (ranging, in early anticipated applications, from micro-electronics to cosmetics!). However, commercial players in, for example, micro-electronics and avionics also possess key capabilities (for

---

[11] In contrast, in a field such as information technology, in which regulation is much less central, there is no clear innovation hierarchy (Tait and Williams, 1999).

example, in process technologies applicable to micro- and nano-scale manipulation). The system of hazard regulation for nanotechnology has not yet been established – and there are signs of an emerging controversy over the health and environmental hazards that may be associated in particular with nano-materials, leading to the recent suggestion that nano-scale materials will require separate safety testing and approval from their macro-molecular counterparts (Royal Society, 2004).

Exploration and Exploitation of Emerging S&T

This view of emergent areas of S&T suggests a rich matrix of options for exploration with many possible avenues for exploitation. For example, the generic nature of developments in NBIC provides the potential for many possible applications.

Emerging areas of S&T typically involve a cluster of complementary developments in which scientific knowledge, instrumentation, laboratory and manufacturing techniques, and end-product innovation are co-created. According to this view, breakthrough S&T does not involve a single development, but rather many related activities, and it is this richness that offers the potential for some successful outcomes. Even if the main goal towards which the S&T activity is orientated proves unrewarding, there are likely to be many "spin-offs" due to the rich nature of the complementary developments.

One potential focus for assessment of breakthrough S&T could therefore hinge on this issue of generic richness and clustering of complementary activities. Is it possible to assess whether some research agendas are narrower than others, and if so, does lower richness mean less opportunities for linkages into exploitation? Should heterogeneity be an explicit aim of research agendas since it may enhance the prospects for cross-fertilisation?

Another issue concerns how to balance concertation with heterogeneity – and whether strategies can be developed that "hedge" (Collingridge, 1992) against the problems that may arise where past development choices take us down roads that prove unproductive or bring undesirable, unforeseen consequences.

Funding Choices: Assessment and Structuring of Emerging S&T

Because of the difficulty in predicting final applications for emerging S&T, assessment must necessarily look more at the process than at the product. Within the period in which funding decisions are being made and evaluated, judgements can only be made with any reliability about the way that research is carried out, since it will not be possible to assess the outcomes, if at all, until many years after the decision. However, the question still remains regarding what aspects of the process can be usefully assessed, and who should do the assessing.

It is well understood that self-assessment by specialist communities is problematic. Researchers with close ties to a particular approach are likely to be too narrowly focused in the criteria they choose for assessment, and overly optimistic with regard to overcoming obstacles (Garud and Ahlstrom, 1997). Moreover, such "insiders" will shape knowledge through the way that they collect and present data on their preferred approach. Typically, disciplinary speciality leads to a narrow focus as regards the choice of what data to collect. Specialists measure what they know how to measure, not necessarily what should be measured. Firms behave in the same way as research teams, following a routine-based approach as described by the organisation theories of Simon and March and more recently by competence-based theory, in evolutionary economics. It would be too expensive and time-consuming to measure and assess all the parameters of a new area of science or technology, so inevitably a research team or firm must take decisions in a context of incomplete information; they must make presumptions and apply heuristics to determine which avenues to pursue. Often this will lead them to focus on their own area of expertise.

Insider assessment can become self-reinforcing and tautological if both are modelled on the same assumptions (Constant, 1987). "Data inconsistent with a particular assessment routine tended to be either ignored, or perceived as noise" (Garud and Ahlstrom, 1997, p.42). Insider assessment involves "enactment cycles" which are self-referential, based on exemplary cases, and geared towards a specific technological trajectory. Outside assessment uses "selection cycles" involving more formalised comparative analysis based on standardised (for example, statistical) data. This requires outsiders to learn about details of the research before they can determine what characteristics should be used for comparison. Likewise, insiders learn about the assessment methods used by outsiders.

If comparisons based on such standards become the benchmark for obtaining funding then they will most likely determine the course of future research, and exclude research that does not conform. The process of determining what measurements provide for valid comparisons is thus critical to the survival of a particular approach. Assessment approaches used for evaluation determine the construction of reality (Latour and Woolgar, 1979) and an approach becomes "black-boxed". A key development in the emergence of new S&T is thus the development of instrumentation and techniques to provide a "level playing field" for comparison and assessment.

However, this black-boxing of new fields of S&T can make them overly "concrete" to policy-makers. Policy-makers and generalist managers of research funding may attribute greater certainty to an emerging area of S&T than those at the heart of its development. They lie in the "certainty trough": committed to the emerging field without fully comprehending the uncertainties involved (MacKenzie, 1998). New fields may also thus gain policy commitment which can sustain the field even when its practitioners have moved on. For example, in the UK the concept of Computer Aided Production Management was coined and became the focus of a joint EPSRC/DTI research and awareness programme at precisely the point that practitioners concluded that this was no longer the optimal focus of attention (Clausen and Williams, 1997).[12]

This brings us back to the central problem. We can identify emerging fields, but assessing their long-term potential to provide breakthrough S&T is more difficult. The fact that fields emerge, and attract funding and policy support, does not in itself validate those fields. It may simply reflect the relative political ability of particular researchers and programme leaders. Given such concerns, to what extent should governments attempt to pick winners as opposed to promoting diversity, which may yield alternative technology trajectories?

Issues in the Assessment of S&T

A general point which must be stressed is that S&T assessment can never be regarded as simply an objective "technical" process. The assessment framework, and the criteria by which performance is measured and judgements are made, are profoundly shaped by the social position, expertise, and perspective of those involved. In this final section of the chapter we consider some different approaches that might be adopted to improve the effectiveness of assessment for breakthrough S&T.

Particularly important is the choice of the unit of assessment. Funding is typically allocated to projects, perhaps as part of broader programmes – and assessment follows this. However, although projects are a logical unit of assessment for funding bodies, they typically do not reflect the way that research is carried out. Knowledge creation (and breakthroughs in S&T) is often associated with a series of projects, and with the intellectual development of an individual, a research team, a laboratory or indeed a network – an invisible college. The assessment of individuals is problematic because it ignores the value of research teams and collaborative relationships. According to Rogers et al. (2001):

> When we asked scientists questions about their "projects", we found that most, and especially the most productive of them, did not conceptualize their work on a project basis. Projects were viewed more as necessary fictions. Money from various funding sources was blended, used strategically, with no tight coordination among "projects". Especially in longer-term "projects", the only bureaucratic footprints often are in the hurriedly written paragraphs submitted for required progress reports.

Most previous attempts at assessment of S&T research have focused on the measurement of socio-economic benefits (ranging from macroeconomic assessments of the overall contribution of R&D to GNP, to microeconomic tracing of the returns generated by specific innovations, and measurements of research

---

[12] Similar comments emerge in the UK today about the identification of 'genomics' as a focus of a socio-economic research programme by the Economic and Social Research Council, in a period in which scientists are emphasising the turbulence and obsolescence of sub-disciplines.

activity such as patents, citation counts, licensing, etc. (Georghiou and Roessner, 2000)). Such assessment exercises are, of course, retrospective, and thus not useful for assessing emerging S&T as it happens.

A more promising alternative to evaluation of economic outcomes (which has been termed a "pre-economic" approach) may be found, for example, in the work of Bozeman and Rogers (2002). Arguing that "innovation and knowledge flows cannot be assessed independently of the collective arrangement of skilled people, their laboratories and instruments, their institutions, and their social networks of communication and collaboration", they have proposed an approach based on assessment of what they call the "knowledge value collective". Accordingly, "it is the collective arrangements and their attendant dynamics, not the accumulated innovation products, that should more properly be considered the main asset when assessing the value of research" (Rogers and Bozeman, 2001). Bozeman and Rogers have further developed this approach in what they term the "churn model" of knowledge. Value is based on the *range and repetition of uses of scientific and technical knowledge*. "Use is an explicit act of valuation on the part of the user. If scientific and technical knowledge is purchased at great expense, use indicates value; if scientific and technical knowledge costs nothing (in monetary terms) use is nonetheless, a direct imputation of value" (Bozeman and Rogers, 2001, p.772).

Roadmapping

Another promising approach to assessing breakthrough S&T is provided by roadmapping, which emerges from attempts at technology forecasting and foresight. Technology forecasting arose in the 1960s as a way of attempting to predict the nature and impacts of technological change (Cronberg, 1996). More recently, some nations, including the UK, have adopted Technology Foresight as a tool aimed at improving national well-being and economic competitiveness. Put simply, Foresight operates by bringing together business and scientists in order to help identify trends, and to assess their significance. Roadmapping differs historically from Technology Forecasting and Foresight in that it was developed within industry, and in particular within the semiconductor industry. Roadmaps were first developed within Motorola in the 1970s as a way of bringing together different parts of the company to develop a consensus vision of how future technology areas might affect the business.[13]

However, the approach has now found wider application across a broad spectrum of users, from research organisations and academia through industry and public sector administration. In some cases, retrospective roadmapping has been used to demonstrate the role of specific investments in S&T, but our main interest

---

[13] According to Bob Galvin of Motorola these roadmaps "communicate visions, attract resources from business and government, stimulate investigations, and monitor progress. They become the inventory of possibilities for a particular field, thus stimulating earlier, more targeted investigations" (Galvin, 1998).

lies in prospective or forward roadmapping which is concerned with guiding future developments (Kostoff and Schaller, 2001).

Because of its practice-orientated origin, and diverse application, roadmapping is not a precisely defined technique. Defined broadly: "A 'roadmap' is an extended look at the future of a chosen field of inquiry composed from the collective knowledge and imagination of the brightest drivers of change in that field" (Galvin, 1998). Typically, "an S&T roadmap provides a consensus view or vision of the future S&T landscape available to decision-makers" (Kostoff and Schaller, 2001).

Roadmaps can be seen as an attempt to make explicit the guiding assumptions within an industry, and in the US semiconductor industry this involved a high degree of collaboration amongst competitors. Their benefits derive from alignment within and between organisations, and the communication that this requires: "An important aspect of the road mapping technique is the multidisciplinary cross-functional working that it requires in order to fulfil its objective of providing common guidance for the whole organisation" (Probert and Shehabuddeen, 1999). However, this coalescence around a shared vision of the future can mean that the technological trajectory becomes self-fulfilling. A linear view of development is perpetuated which can then exclude alternative approaches (Walsh, 2004).

However, roadmapping can also be "used as a means for assessing the impact of potentially disruptive technologies and markets on business plans and systems (Phaal et al., 2004). Walsh (2004) describes attempts to develop a roadmap for MEMS (a set of technologies, variously labelled as microsystems, micromachining, and top-down nanotechnology) which highlighted some of the key challenges. In particular, in an emerging area such as MEMS it is difficult, if not impossible, to identify the likely product and hard to specify which will be the major technology drivers; there was no shared terminology, and there were many potential technology trajectories, a vast number of potential applications, and a complete lack of industry cohesion as regards standards and objectives (Walsh, 2004). Eventually, however, a MEMS roadmap was developed, taking these characteristics of emerging technology into account, identifying major technology pathways and product platforms, and specifying as far as possible a common terminology which could form the basis for the development of new standards, and providing a detailed account of a range of interrelated technologies, products and markets, enabling the identification of "the potential next 'big commercial opportunity', bottlenecks, and roadblocks" (Walsh, 2004).

Roadmapping can achieve two key things in guiding emerging and breakthrough S&T. First, the process of developing a roadmap involves bringing together a wide range of relevant experts, which, depending on how they are chosen, has the potential to facilitate interdisciplinary linkages that might not otherwise have occurred. Second, this cross-cutting or "churning" of research activities can be institutionalised in the roadmap.

The process of roadmapping can be used for assessment. Roadmapping brings together a range of relevant experts, covering a diverse range of specialities from both user and researcher perspectives. Together these experts can produce a roadmap more comprehensive, and more sensitive to emerging S&T and potential linkages between differing areas, than any single perspective would provide. The resultant roadmap thus provides an assessment of where the field is currently at, and what are the range of potential future developments, both incremental and breakthrough. This roadmap, and subsequent revisions, then provide an ongoing process of assessing how the field is developing. There are issues about the choice of the scope of the roadmap and of who constitutes a "relevant expert", which are not always evident in advance and so roadmapping should be an iterative process. As the roadmap is developed, linkages will be made which may require input from a field not previously recognised as significant (Kostoff and Schaller, 2001).

Finally, an explicit focus on disruptive S&T in roadmapping can help commercial organisations to overcome their inherent bias towards sustaining technologies which fit into existing production and marketing systems, and which promise short-term increases in profitability (Kostoff et al., 2004). Roadmaps, both industry-wide and firm-orientated, could thus institutionalise mechanisms to highlight the potential of longer-term disruptive technologies, and negate the tendency of individuals and business units to focus on short-terms goals. In particular, roadmapping can be used to help provide a longer-term strategic perspective.

Generic "Richness"

We have already noted how radical innovations have been characterised by a clustering of innovations in both "core" capabilities and a range of ancillary innovations, while systemic innovations have been linked across a range of different application domains. Arie Rip and co-workers at the University of Twente have suggested that these two features could be used to assess the potential of emerging research fields in terms of their generic richness conceived in two ways. First, richness in relation to a diverse cluster of knowledge, instrumentation, and techniques which emerge together as each new advance stimulates or requires progress elsewhere (compare Rosenberg (1976) "focussing" and Hughes (1987) "reverse salients"), in the development of a new field of S&T. Second, richness related to the level of linkages that an emerging field of research has across other disciplines.

The challenge is how to assess this richness. Two aspects seem amenable to assessment: organisational structure and knowledge flows. Organisational structures can encourage or limit the cross-fertilisation that builds and exploits S&T richness, both within and between organisations. Organisational procedures can be adopted to enhance cross-disciplinary links, networking and cross-fertilisation, ensuring that managers with market knowledge and sufficient power to promote a new idea come into regular contact with researchers developing potential breakthrough S&T (O'Connor and Rice, 2001). At the opposite extreme, organisational structures and management systems that have rigid vertical hierarchies and little potential for lateral linkages can be seen to stifle the cross-fertilisation that is essential to the generation and exploitation of richness in new

S&T (Kolodny et al., 1996). Cross-cutting linkages also increase the opportunities for emerging S&T to be applied to a range of potential applications. Even where research is strongly focused on a particular application, the existence of appropriate linkages increases the likelihood that other, unforeseen, usages can be realised. In some cases, new S&T may prove unsuitable for its original intended application, but may prove fruitful elsewhere (Kirchhoff and Walsh, 2000).

Conclusions

This chapter has noted the challenges for science and technology governance and policy-making that may be associated with discontinuities in development pathways – captured, for example, by concepts such as "breakthrough" science and technology and disruptive technologies – and has identified a number of responses to the consequent problems in identifying and assessing breakthrough S&T proposals, including the use of roadmaps and assessments of linkages and richness of S&T fields. These techniques of course do not resolve the problems – for example, the risks that undue alignment and concertation of effort may divert attention from promising new lines of exploration. However, they can act as systematic sensitising devices and make explicit the kinds of consideration that need to be balanced (alignment versus diversity; generic versus specific, etc.). They also point to a broader shift in the governance of science and technology. Roadmapping, and the more advanced practices emerging from Foresight initiatives, went beyond earlier concerns to anticipate technology trajectories, and instead involved diverse actors (research, industry, policy, lay) in discussing the choices surrounding technological development. Subjects are integrated in the policy process – indeed the key outcomes arguably are as much the *process* of interaction and the emergence of new relationships, as the substantive findings. Though there may be debates about the range of players and voices included, and how best they may be combined, these developments can be seen as exemplary of the wider shift from policy-making to governance that this book explores.

The chapter also flags the potential contribution of socio-economic research in charting historical processes of S&T breakthroughs, and in providing exemplars and tools that can guide strategies and tools to identify and pursue new breakthroughs as they emerge. We must also consider, however, the possibility that further new developments might follow rather different patterns (and thus require rather different policy tools) than the past. There are obvious risks of seeking to conceive future rounds of innovation too narrowly around historical experiences. For example, widespread public concern about the potential health and environmental hazards of genetically-modified organisms has framed subsequent discussions of other life science technologies and is now being used as a template for discussing nanotechnology (Grove-White et al., 2004). However, the attempt to draw lessons from particular cases may bring perils as well as insights (Phoenix and Drexler, 2004). Care and caution are needed in drawing lessons. Here current approaches within Science and Technology Studies emphasise the contingency of outcomes, depending upon interactions amongst complex configurations of actors and the particular characteristics of a technology and the contexts of its development and use (Russell and Williams, 2002). Mechanistic extrapolation will not succeed. But a growing body of empirical studies can help us map the main dilemmas and parameters surrounding decisions, and the strategies for reconciling competing considerations.

References

Anderson, P. and Tushman, M. L. (1990), 'Technological Discontinuities and Dominant Designs: A Cyclical Model of Technological Change', *Administrative Science Quarterly*, Vol. **35**, pp.604-633.

Arrow, K. (1962), 'The economic implications of learning by doing', *Review of Economic Studies*, Vol. **29**, pp.155-173.

Bozeman, B. and Rogers, J.D. (2002), 'A churn model of scientific knowledge value: Internet researchers as a knowledge value collective', *Research Policy*, Vol. **31**, pp.769-794.

Braun, E. and MacDonald, S. (1982), *Revolution in Miniature: History and Impact of Semiconductor Electronics*, Cambridge: Cambridge University Press.

Brosveet, J. and Sørensen, K.H. (2000), 'Fishing for fun and profit: National Domestication of Multimedia: The Case of Norway', *The Information Society*, Vol. **16**(4), pp.263-276.

Clausen, C. and Williams, R. (eds) (1997), 'The Social Shaping of Computer-Aided Production Management and Computer Integrated Manufacture', Vol. **5**, Proceedings of International Conference, COST A4, Luxembourg: European Commission DGXIII.

Collingridge, D. (1992), *The Management of Scale: Big Organizations, Big Decisions, Big Mistakes*, London: Intl Thomson Business Press.

Constant, E. W. (1987), 'The Social Locus of Technological Practice: Community, System or Organization?' in Bijker, W.E., Hughes, T.P. and Pinch, T. (eds), *The Social Construction of Technological Systems*, Cambridge: MIT Press.

Cronberg, T. (1996), 'Technology Assessment in the Danish Socio-Political Context', *International Journal of Technology Management*. Vol. **11**(5/6).

David, P. (1985), 'Clio and the economics of QWERTY', *Economic History*, Vol. **75**, pp.332-337.

Fleck, J. (1982), 'Development and Establishment in Artificial Intelligence', in D. Reidel (ed.), *Scientific Establishments and Hierarchies*, Vol. **6**, pp.169-217 of 'Sociology of the Sciences', Dordrecht: Norbert Elias, Publishing Company.

Freeman, C. (1974), *The Economics of Innovation*, Harmondsworth: Penguin.

Galvin, R. (1998), 'Science Roadmaps', *Science*, Vol. **280**, 8 May, 803.

Garud, R. and Ahlstrom, D. (1997), 'Technology assessment: a socio-cognitive perspective', *Journal of Engineering and Technology Management*, **14**, pp.25-48.

Georghiou, L. and Roessner, D. (2000), 'Evaluating technology programs: tools and methods', *Research Policy*, Vol. **29**, pp.657-678.

Gibbons, M., Limoges, C., Nowotny, H., Schwartzman, S., Scott, P. and Trow, M. (1994), *The New Production of Knowledge. The Dynamics of Science and Research in Contemporary Societies*, London: Sage.

Gilfillan, S. (1970), *The Sociology of Invention*, Cambridge, MA: MIT Press.

Grove-White, R., Kearnes, M., Miller, P., Macnaghten, P., Wilsdon, J. and Wynne, B. (2004), 'Bio – to – Nano? Learning the Lessons, Interrogating the Comparison', A

Working Paper by the Institute for Environment, Philosophy and Public Policy, Lancaster University and Demos.

Hughes, T.P. (1987), 'The evolution of large technological systems', in Bijker, W., Hughes, T.P. and Pinch, T. (eds), *The Social Construction of Technological Systems*, Cambridge: MIT Press, pp.51-82.

Kassicieh, S.K., Walsh, S.T., Cummings, J.C., McWhorter, P.J., Romig, A.D. and Williams, W.D. (2002), 'Factors differentiating the commercialization of disruptive and sustaining technologies', *IEEE Transactions on Engineering Management*, Vol. 49(4), November, pp.375-387.

Kirchhoff, B. and Walsh, S. (2000), 'Entrepreneurship's role in commercialization of disruptive technologies', *Unternehmer und Unternehmensperpektive fur Klien-und-Mittelunternehman*, Berlin, pp.323-332.

Kolodny, H., Liu, M., Stymne, B. and Denis, H. (1996), 'New technology and the emerging organisational paradigm', *Human Relations*, Vol. 49(12), pp.1457-1487.

Kostoff, R.N. and Schaller, R.R. (2001), 'Science and Technology Roadmaps', *IEEE Transactions on Engineering Management* Vol. 48(2), May, pp.132-143.

Kostoff, R.N., Boylan, R. and Simons, G.R. (2004), 'Disruptive technology roadmaps', *Technological Forecasting and Social Change*, Vol. 7, pp.141-159.

Kuhn, T. (1962), *The Structure of Scientific Revolutions*, University of Chicago Press.

Latour, B. and Woolgar, S. (1979), *Laboratory Life: The Social Construction of Scientific Facts*, London: Sage.

Law, J. (1988), 'The Anatomy of a Socio-Technical Struggle: The Design of the TSR 2', in Elliott, B. (ed.), *Technology and Social Process*, Edinburgh: Edinburgh University Press, pp.44-69.

MacKenzie, D. (1998), 'The Certainty Trough', in Williams, R., Faulkner, W. and Fleck, J. (eds), *Exploring Expertise*, Basingstoke: Macmillan, Chap 15, pp.325-329.

National Research Council (1999), 'Funding a Revolution: Government Support for Computing Research Committee on Innovations in Computing and Communications', *Lessons from History*, Washington, D.C: National Academy Press.

National Science Foundation (2002), 'Converging Technologies for Improving Human Performance', Arlington, VA: NSF June 2002. The report is available in a draft version on the Internet at: http://www.itri.loyola.edu/ConvergingTechnologies.

Nowotny, H., Scott, P. and Gibbons, M. (2001), *Re-Thinking Science: Knowledge and the Public in an Age of Uncertainty*, London: Polity Press.

Oakley, B. (1990), 'Look back in Alvey: why support for R&D is not enough', *Edinburgh PICT Working Paper 18*, Edinburgh: Edinburgh University.

O'Connor, G. C. and Rice, M.P. (2001), 'Opportunity Recognition and Breakthrough Innovation in Large Established Firms', *California Management Review*, Vol. 43, No 2, pp.95-116.

Phaal, R., Farrukh, C.J.P. and Probert, D.R. (2004), 'Technology roadmapping – A planning framework for evolution and revolution', *Technological Forecasting & Social Change*, Vol. 71, pp.5-26.

Phoenix, C. and Drexler, E. (2004), 'Safe exponential manufacturing', *Nanotechnology* No. 15, pp.869-872.

Prahalad, C. K. and Hamel, G. (1990), 'The core competence of the corporation', *Harvard Business Review*, Vol. 68(3), May/June, pp.79-91.

Probert, D. and Shehabuddeen, N. (1999), 'Technology road mapping: The issues of managing technology change', *International Journal of Technology Management*, Vol. 17(6), pp.646-661.

Rammert, W., et al. (eds) (2003), INNOVERSITY Innovation and Diversity in a Knowledge Society: Towards a New Regime of Distributed Innovation? Proceedings of international research workshop, 18-20 September 2003, European Academy Berlin, Berlin: Technical University of Berlin.

Rip, A. (2002), 'Science for the 21st Century', in Tindemans, P., Verrijn-Stuart, A. and Visser, R. (eds), *The Future of the Sciences and Humanities*, Amsterdam University Press.

Rogers, J.D. and Bozeman, B. (2001), 'Knowledge Value Alliances: An Alternative to the R&D Project Focus in Evaluation', *Science, Technology & Human Values* Vol. **26**(1), pp.23-55.

Rogers, J.D., Bozeman, B. and Chompalov, I. (2001), 'Obstacles and opportunities in the application of network analysis to the evaluation of R&D', *Research Evaluation*, Vol. **10**(3) December, pp.161-172.

Rosenberg, N. (1976), *Perspectives on Technology*, Cambridge: Cambridge University Press.

Royal Society (2004), *Nanoscience and Nanotechnologies: Opportunities and Uncertainties*, London: Royal Society.

Russell, S. and Williams, R. (2002), 'Concepts, Spaces and Tools for Action? Exploring the Policy Potential of the Social Shaping of Technology: Perspective', in Sørensen, K.H. and Williams, R. (eds) *Shaping Technology, Guiding Policy: Concepts, Spaces and Tools*, Aldershot: Edward Elgar, Chapter 4, pp.133-154.

Schumpeter, J. (1942), *Capitalism, Socialism, and Democracy*, New York: Harper & Brothers.

Tait, J. and Williams, R. (1999), 'Policy Approaches to Research and Development: Foresight, Framework and Competitiveness' *Science and Public Policy*, 1999, Vol. **26**(2), pp.101-112.

van Lente, H. and Rip, A. (1988), 'Expectations in Technological Developments: An Example of Prospective Structures to be Filled in by Agency', in Disco, C. and van der Meulen, B. (eds), *Getting New Technologies Together: Studies in Making Sociotechnical Order*, Walter de Gruyter, pp.203-229.

Walsh, S.T. (2004), 'Roadmapping a disruptive technology: A case study. The emerging microsystems and top-down nanosystems industry', *Technological Forecasting & Social Change*, Vol. **71**, pp.161-85.

Walsh, S., Kirchoff, B. and Newbert, S. (2002), 'Differentiating market strategies for disruptive technologies', *IEEE Trans. Eng. Manage.*, Vol. **49**(4), pp.341-351.

PART II
DEVELOPING AN INTEGRATED
POLICY APPROACH

Chapter 4

Life Science Innovation: Policy and Foresight

Thomas Reiss and Joyce Tait

Introduction

Public policy is a key factor influencing innovation in the major sectors of biotechnology application (biopharmaceuticals, agro-food and equipment and supplies) (Senker et al., 2001; Tait et al., 2004). In the 1990s most, if not all, of the European national policy systems intensified their efforts in supporting the development of the biotechnology knowledge base and its industrial application (Enzing et al., 1999). However, life science innovation is an area where there is a particular need to integrate the societal with the scientific and technical aspects of innovation.

This chapter considers the performance of European countries in achieving their aims for promotion of science, technology and innovation (STI) in life sciences, along with a range of targeted and generic policy instruments adopted to support these aims. We then examine a range of Foresight studies and their potential role as a mechanism for integration of STI-related policies. The more sophisticated of these Foresight initiatives, which incorporate a broad range of societal concerns, could begin to develop as mechanisms to illustrate how life science innovation will play a role in determining societal futures and also how scientific, technical and social aspects of life science innovation will interact with one another to determine the future shape of these industry sectors.

To achieve effective, integrated governance of the life sciences, policy-makers will need to strike an appropriate balance between promotion and regulation of innovation; between what is feasible technically and commercially and what is publicly acceptable and desirable; and to deliver the claimed societal benefits without challenging accepted societal norms.

Performance of European Countries in Biotechnology

Despite a widespread and increasing policy interest in biotechnology in Europe (European Commission, 2002), there is a surprising lack of systematic, internationally comparable data on the performance of national or sectoral biotechnology innovation systems (Reiss et al., 2003[1]; European Commission, 2003).

Recent national studies include an analysis of the development of the biotechnology industry in France, identifying different business models of biotechnology firms (Mangematin et al., 2003). A Swedish Agency for Innovation Systems (VINNOVA) analysis of biotechnology performance using science, technology and economic indicators (Sandstrom and Norgren, 2003) found that Sweden performed well in various sub-areas of life sciences between 1986 and 2001 as measured by publications per capita, achieving rankings among the top three countries in the world. Another Swedish study focused on formal knowledge collaboration in the bio-pharmaceutical sector and the relative importance of co-location for formal knowledge collaboration (McKelvey et al., 2003), concluding that co-location is more important for collaboration between firms and universities compared to inter-firm collaborations, pointing to the significance of the local knowledge base for the commercial development of biotechnology. An evaluation of Finnish biotechnology firms has compiled a broad set of commercialisation indicators (Hermans and Luukkonen, 2002). Commercial development and academic activities in the life sciences in the Netherlands were analysed by Enzing et al. (2002), showing that commercial development of biotechnology in the Netherlands occurred at a slower rate compared to other European countries.

The European biotechnology innovation scoreboard (European Commission, 2003) is the most recent attempt to fill the information gap on biotechnology performance. It highlights strengths and weaknesses of EU member states in biotechnology innovation and identifies Denmark, Sweden and Finland as the leading EU performers in biotechnology. A background report on European competitiveness (European Commission, 2001) derived from a database which is not accessible to the public (Allansdottir et al., 2002) confirms the leading position of the US in innovative activities related to biotechnology and also points out that an inadequate supply of cutting-edge scientific research may pose problems for the further development of the biotechnology industry in Europe. A consultancy company also provides yearly reports on the European biotechnology industry with a strong focus on firm creation and economic performance (Ernst and Young, 2001) but it is difficult to assess such surveys because data accessibility is limited.

The overall picture presents Europe as a very diverse area with strong research activities in some life science fields and weaknesses related to the exploitation of the biotechnology research base – a situation referred to as the European paradox (European Commission, 1996). The large countries are usually considered to be high performing and recent studies, as noted above, also point to a good performance by the Nordic countries.

---

[1] The project '*Effectiveness of innovation policies in high technology sectors in Europe (EPOHITE)*', contract no. HPV1-CT-2001-00005 was carried out by a research team from Fraunhofer ISI (Germany) as co-ordinator, TNO Institute for Strategy, Technology and Policy (Netherlands), SPRU (United Kingdom), and UMR GAEL (France) between April 2001 and July 2003; see http://www.epohite.fhg.de/.

The EPOHITE analysis (Reiss et al., 2003) indicates that there is an increasing significance of biotechnology related research in all European member states leading to an increasing output of scientific publications in quantitative and qualitative terms as measured by publications per capita, citation rates and the share of basic research publications. The overall performance of European member states in providing a knowledge base for biotechnology (Table 4.1) reveals the following ranking: the three Nordic countries Sweden, Denmark and Finland present the best performance, followed by the United Kingdom, the Netherlands, Belgium and Austria, all performing above the European average. France, Germany, Ireland, Spain, Italy, Portugal and Greece comprise, in this order, the rest of the ranking list, performing below the European median.

Table 4.1   Index values of biotechnology (BT) knowledge base indicators[2]

	BT publications per million capita		Citations per BT publication		Basic BT publications per total BT publications		Average knowledge base Indicator	
	1995/ 1996	1999/ 2000	1995/ 1996	1999/ 2000	1995/ 1996	1999/ 2000	1995/ 1996	1999/ 2000
Austria	6.78	7.03	7.78	8.50	6.32	7.41	6.96	7.65
Belgium	7.92	8.49	8.62	7.44	7.94	7.28	8.16	7.74
Denmark	11.49	11.83	6.95	8.26	6.76	7.39	8.40	9.16
Finland	11.50	11.10	8.30	7.65	7.03	7.21	8.94	8.65
France	6.34	6.01	6.74	7.36	8.36	8.64	7.15	7.34
Germany	5.66	5.77	7.59	7.64	8.56	8.11	7.27	7.17
Greece	1.92	2.42	5.67	4.59	5.74	4.99	4.44	4.00
Ireland	5.06	5.48	6.02	7.62	4.45	5.45	5.74	6.18
Italy	4.02	4.08	7.03	6.49	6.16	6.58	5.74	5.72
Netherlands	11.58	9.85	7.80	7.90	7.86	6.31	9.08	8.02
Portugal	1.33	1.87	5.59	5.25	7.51	7.92	4.81	5.01
Spain	3.54	4.04	4.73	5.59	7.96	7.73	5.41	5.79
United Kingdom	8.55	8.22	9.14	8.47	8.23	7.76	8.64	8.15
European median							7.14	7.49

Comparing 1999/2000 with the mid-1990s, Denmark, Austria, Ireland and Spain improved their position, while the Netherlands, Germany and Italy moved down the ranking list. All other countries kept their positions. These EPOHITE results are supported by the European biotechnology innovation scoreboard

---

[2] In order to compare the performance of the different countries, a scaling system was used which transfers the scores of each indicator to a 100-point scale. 100 points represent the sum of the indicator values of all countries. For comparison, the median value of the 100-point scores is calculated. An overall performance indicator was calculated as the average of the individual indicators.

(European Commission, 2003) which also found that Sweden, Denmark and Finland followed by the United Kingdom, the Netherlands and Belgium produced the highest output of scientific publications in biotechnology in 2000, while the Mediterranean countries occupied a rather weak position. Interestingly, Switzerland produces the highest number of biotechnology publications related to population among all European countries and also compared to the United States and Canada, which are ranked between the Netherlands and Belgium.

Most member states had a similar specialisation pattern, in that pharmaceutical biotechnology comprised almost 50 per cent of publications, other important areas being plant biotechnology and biotechnology related to the exploration of the cell factory. Between 1995 and 2000 there were only minor shifts between the different areas with the exception of plant biotechnology which considerably lost its significance. During this period the legal environment for plant biotechnology became more difficult due to the de facto moratorium on the authorisation for marketing genetically modified (GM) crops issued in June 1999. L'Heureux et al. (2003) found that almost 39 per cent of 165 European institutions (large firms, small and medium-sized enterprises, public sector research organisations and universities) doing research on GM crops cancelled at least one such project between 1998 and the beginning of 2002, a main reason being the unclear legal situation. Universities and research institutions also cited limited financial resources as a main constraint. These findings, along with the EPOHITE analysis, suggest that the unclear legal situation led to a reduction in research activities in plant biotechnology, reflected in decreasing scientific output. By the time the legal environment has become more stable and/or more favourable for commercialisation of GM crop research, the knowledge base may be less able to provide the required know-how.

The analysis of commercial performance of European member states (Table 4.2) reveals that the biotechnology industry has grown considerably in the period 1996 to 2000, the strongest growth being in Germany and the Nordic countries, but that the industry is still at a rather early stage. Measuring commercialisation performance of European countries by the amount of invested venture capital into biotechnology per population reveals an increasing flow of venture capital into almost all European countries during the 1990s. The number of initial public offerings (IPOs) placed in the periods 1995 to 1997 and 1998 to 2000 gives an indication of the development stages of the biotechnology industry in the member states. Accordingly, Denmark, the United Kingdom and Germany are the countries with the most mature industries. As measured by patent applications, Denmark shows by far the best performance followed by Belgium, the Netherlands, Sweden, Germany and the United Kingdom, all above the European average. Finland, Austria, France, Ireland and the Mediterranean countries (in this order) form the second half of the ranking list. Similar results were obtained by the biotechnology innovation scoreboard (European Commission, 2003).

Table 4.2 Index values of biotechnology (BT) commercialisation indicators used for performance clustering[3]

	Venture capital per million capita		BT companies per million capita		BT patents per million capita		IPO per million capita		Average commercialisation indicator	
	1995/1996	1999/2000	1995/1996	2000	1995/1996	1999/2000	1995-1997	1998-2000	1995/1996	1999/2000
Austria	0.00	0.24	n.a.	n.a.	6.74	6.74	0.00	6.55	2.25	4.51
Belgium	23.18	15.92	8.75	8.05	10.94	13.02	7.07	0.00	12.49	9.25
Denmark	6.54	21.94	13.52	15.27	23.08	20.51	27.00	30.07	17.53	21.95
Finland	5.66	10.01	14.87	18.88	9.36	6.84	0.00	10.30	7.47	11.51
France	7.74	8.07	4.39	3.80	5.55	6.08	6.14	2.70	5.96	5.16
Germany	7.90	18.90	3.23	5.04	7.69	9.30	1.75	9.63	5.14	10.72
Greece	0.00	0.00	n.a.	n.a.	0.16	0.35	0.00	0.00	0.05	0.12
Ireland	10.68	0.21	18.21	10.86	3.97	5.23	19.72	0.00	13.14	4.07
Italy	0.42	1.20	1.42	1.15	1.43	1.31	2.50	1.85	1.44	1.38
Netherlands	13.99	9.69	8.18	6.80	12.69	11.31	0.00	13.46	8.71	10.32
Portugal	2.67	0.26	n.a.	n.a.	0.08	0.50	0.00	0.00	0.92	0.25
Spain	0.01	0.85	0.97	0.80	0.87	1.01	0.00	0.00	0.46	0.66
United Kingdom	20.84	8.05	7.79	5.89	7.85	7.44	19.55	13.49	14.01	8.72
European median									6.71	6.94

---

[3] See footnote to Table 4.1 for calculation of indicators.

Combining the performance rankings for creating a knowledge base for biotechnology and for commercialisation of biotechnology (Tables 4.1 and 4.2) indicates that there is a close relationship between scientific and commercialisation performance. Most countries perform either weakly or strongly in both categories. The exceptions are Germany and Austria. Austria has a strong knowledge base performance, while commercialisation performance is weak. Germany is strong in commercialisation and below average in knowledge base related performance. Combining both types of performance identified four clusters, ranked according to overall performance: (1) Denmark, Sweden and Finland; (2) the Netherlands, the United Kingdom, Belgium and Germany; (3) Austria, France and Ireland; and (4) Italy, Spain, Portugal and Greece.

An array of local, national and international factors is responsible for these apparent differences in the performance of life science industry sectors across EU regions and between the EU and the United States. The following sections explore the impact of a range of policies and other governance-related influences.

Policies Promoting Biotechnology[4]

The strong relationship between scientific and commercial performance in biotechnology implies that it is important to consider the whole innovation process and take a systemic perspective when designing and evaluating policy instruments for supporting biotechnology.

Policies are part of an interdependent system of forces that influence the innovation process, and the outputs of policy instruments may interact with other measures. Additionally, policies may have an impact on one sector or technology without targeting it explicitly. We consider these factors when identifying policy instruments influencing the development of biotechnology, its diffusion and its industrial application. The EPOHITE project differentiated between targeted instruments that have an impact on these sectors without being biotechnology specific (Table 4.3).

We define targeted policy instruments as those which aim directly to promote the development or commercialisation of biotechnology through the creation of the biotechnology knowledge base, encouraging networking activities among the actors involved in the innovation process or facilitating the industrial application of biotechnology. Other targeted policy instruments have a socio-economic dimension dealing with public awareness or ethical aspects related to specific areas of biotechnology research.

Generic policy instruments do not specifically promote biotechnology but contribute to its development and commercialisation indirectly, by promoting scientific and technological development per se, improving the conditions for innovation in general, protecting intellectual property rights or assuring the availability of financial capital in high growth sectors.

---

[4] This section is based in part on Dominguez Lacasa et al. (2004).

Table 4.3   Policy instruments to support life science innovation in Europe between 1994 and 2001

Targeted policy instruments	Generic policy instruments
Policy instruments supporting the biotechnology knowledge base	General science and technology policies
Policy instruments supporting diffusion and industrial application of biotechnology	Legislation on intellectual property rights (IPRs)
	Measures to assure the availability of financial capital in high growth sectors

Targeted Policy Instruments

The importance of developing and renewing the biotechnology science base reflects the significance of scientific knowledge for sustainable innovation in biotechnology based sectors. The EPOHITE analysis of policy instruments targeting the science base from 1994-2001 concentrates on the extent to which policy systems support the development of the biotechnology knowledge base, the mechanisms for the allocation of public funding and the types of policy instruments implemented to support research across Europe.

Belgium, Finland and the UK present the largest per capita expenditure on biotechnology research compared to the relatively low expenditure by Austria, Ireland, Spain, Italy and Portugal. However, large budgets do not seem to be enough to deliver successful results in the production of knowledge. According to the data on scientific performance presented by Reiss et al. (2003), of the three top-spending countries, only Finland is among the top scientific performers.

Three types of mechanism for public funding allocation were identified:

- block grants to research institutes, which in some cases are responsible for developing their own internal programmes;
- response-mode competitive funding;
- programme-oriented competitive funding.

Block grants have been the traditional mechanisms in Spain, Italy and France, where institutes have autonomy to control their research agendas. However, this mechanism seems to provide only limited incentives to stimulate research performance. Block grants may also lead to duplication of research. On the other hand, competitive funding is a flexible system benefiting from peer review procedures. Most countries, except for Italy, are increasingly introducing competitive features into their funding allocation systems, although differences in the degree of competition in national funding systems are important. Finland, Sweden, the United Kingdom and (to some extent) Germany allocate their funds mostly through response-mode competitive mechanisms. Competition between actors appears to be a more effective method to achieve high scientific performance than direct control of funds by research institutions, facilitating the development of unconventional scientific research projects which are not in the strategic agenda of government bodies. Countries such as Finland, Denmark, Germany and the Netherlands which have combined a programme orientation (where the relevant research fields are identified) with peer-reviewed competitive grants have been better able to reach a coherent strategy for research and for enhancing the co-ordination of policy actors.

Countries supporting the development of the knowledge base such as Finland, Sweden, Denmark, Germany and the UK have used both targeted and generic policy instruments to support research and there has been a shift over time from targeted to generic policy instruments. For countries with a strong biotechnology performance like Sweden, Finland or Denmark it may be appropriate to make this shift. However, countries with a poor performance such as Greece, Italy and Spain have provided either low or no support through targeted policies. In Greece, generic policy instruments have been quite strong but the scientific performance low. Generic instruments may be ineffective at supporting the biotechnology knowledge base unless they are preceded by targeted instruments that create a critical mass of biotechnology knowledge base and strengthen national scientific capabilities in the relevant fields. Finland, for example, implemented targeted instruments in the 1990s to concentrate its resources for basic research in the so-called biocentres and introduced a growing number of biotechnology-specific programmes.

Analysis of policy instruments implemented since 1994 reveals a recent shift in public promotion programmes, away from funding fundamental research towards the promotion of more applied and industry-driven research. As a response to the general concern about the development lag in the European biotechnology industry compared to its American counterpart, policy instruments have aimed to promote the industrial application of biotechnology in European member states. In the short term this strategy may stimulate the growth of European industry. However, in the long-term, it could weaken the scientific knowledge pool which is needed to keep up with the rapidly expanding scientific knowledge frontier. As biotechnology is based on science and drives a science-based commercial sector, failing to invest sufficiently in fundamental research so as to renew the knowledge base could be detrimental to European industry.

European Union member states have designed a wide variety of policy instruments to encourage the industrial application of biotechnology (Enzing et al., 2004). However, many of the measures which provide such support are generic rather than targeted, operating at the level of support for innovative industry sectors in general.

An important instrument to promote the commercialisation of life science research is to adapt investment in research to those areas which are relevant to strong national economic sectors. One measure in this direction has been to involve industry in science policy formulation. This can raise industrial awareness of the potential benefits of research, knowledge of the work of national researchers and research on topics of relevance to national industry (Senker et al., 2001). Denmark, Sweden, Finland, Germany, the United Kingdom and France all encourage biotechnology industry involvement in the design of public initiatives. In Portugal, Greece or Spain, industry has no chance to influence policy formulation.

Generic Policy Instruments

At the generic level, the appropriate definition and protection of intellectual property rights for inventions is essential to create incentives for the development of biotechnology. The right to the exclusive use of technical innovations for a limited period of time is said to promote innovation and industry development by encouraging firms to invest in R&D. Traditionally, the main problem in patenting all inventions in Europe has been the lack of a unitary patent providing protection across European countries and this has affected biotechnology at least as much as other areas of innovation. As long as no unitary patent is available, biotechnological inventions are subject to different legislative regimes for patent protection in several countries. This, together with the application costs for the European Patent Office and the complexity of the filing and validation procedures, makes the patent application process inflexible and expensive, discouraging patenting particularly by academic scientists and small biotechnology companies. In 1988 the European Commission (EC) adopted Directive 98/44 on the Legal Protection of Biotechnological Inventions aimed at harmonising national laws in this area, but despite the 30 July 2000 deadline, by October 2003 Germany, Austria, Belgium, France, Italy, Luxembourg and the Netherlands had still not implemented the Directive.

Most, if not all, European countries have strongly promoted the creation of institutions to support the patenting of university research results as part of the transfer of knowledge and technology. Technology transfer organisations (TTOs) can be found in most universities and many science parks. They are responsible for providing support in IPR matters to academic scientists and they also act as bridging institutions between industry and academia. According to Reiss et al. (2003) the view of these technology transfer activities has shifted from being in the area of managing research agreements with firms to one in which the primary task of technology transfer is to assess and protect IP and make it available to industry.

European countries have also implemented policy instruments to promote firm creation and sustainable growth of small and medium enterprises (SMEs). The contribution of SMEs and new technology based firms (NTBFs) to the development of science-based industries like biotechnology, and especially their role in generating labour opportunities and commercialising scientific advances, is a matter of intense discussion in the innovation literature (Storey and Tether, 1998). The rationale behind government support for SMEs and NTBFs has been mainly their contribution to the process of sustaining and increasing variety in capabilities and technologies (Carlsson and Jacobsson, 1997, p.310). A case could certainly be made that government funding for the development of SMEs acts mainly to support the innovation strategies of multinational companies in that, in life sciences at least, it funds the early stages of the transfer of new scientific knowledge to the market place only to have the relevant companies taken over by MNCs once viability of the concept has been demonstrated (Tait and Williams, 1999).

National governments have supported firm creation and the development of SMEs by creating science and technology parks, by implementing public promotion programmes for network formation and collaborative research between NTBFs and PSROs, and through financial support to NTBFs, for example through industrial grants. These have been mostly generic (not biotechnology specific) instruments but in some cases, as in Germany, Belgium and Scotland (for example, Scottish Enterprise, 2001) instruments focusing on the development of the biotechnology industry (biotechnology dedicated instruments) have been a key element of innovation policy. The results of the exploration of policy instruments carried out by Reiss et al. (2003) suggest that in Finland, Denmark, Germany, Ireland and Greece instruments aimed at supporting new technology based firms have had a strong weight in their policy strategies. However, instruments to support innovation activities of companies in more advanced stages of development generally have a low priority or do not exist.

Generic policy instruments promoting capital investment in high-tech sectors have been addressed with the aim of creating favourable financial conditions for industrial development in general. Uncertainty and the high development costs of biotechnology innovations make it difficult for biotechnology companies to acquire the necessary financial capital for their activities. Government grants, credit markets, venture capital markets and stock exchange markets are the major sources for funding in the biotechnology industry. Measures to stimulate innovation activities try to encourage capital investment in high growth sectors and start-ups and SMEs can benefit from these measures.

There are major differences in the importance of policy instruments to promote capital investment in high-tech sectors in national policy strategies. Denmark, Germany, Finland, Belgium and Greece have a strong commitment to promote capital investment in high-tech sectors. The UK, Italy, Austria and Portugal have not recognised sufficiently the need for public initiatives to improve financial framework conditions.

Governments have also tried to stimulate economic activity through involvement in credit transactions. For instance, the German experience shows how soft-loan programmes for start-up companies can have a great influence on the rate of new company formation (Ward, 2001, p.A1). These soft-loan programmes have been carried out by independent government-backed banks who offer loans at favourable interest rates for start-up companies, with the condition that the company has also acquired venture capital investment (matching funds). Start-ups can apply for funding through soft loans from a government-backed bank.

Venture capital is also important in providing equity financing to private companies with rapid growth potential. Venture capitalists have at least three options to realise their return on investment: when the start-up starts making profits, when the start-up is mature enough to go for an initial public offering (IPO) or when the start-up is acquired by another company. Government-backed banks or public institutions for industrial development may act as venture capitalists for biotechnology start-up companies particularly in countries where risk aversion is so high that private venture capital tends to finance safer, more mature and less dynamic companies. In general terms, the availability of venture capital in Europe has improved considerably since the mid-1990s. According to Howell et al. (2003) economic institutions and changes in the entrepreneurial climate have very much supported this trend.

Other policy instruments to encourage equity financing in the start-up phase provide mechanisms that open the possibility to invest in private equity and facilitate the return on investment of venture capitalists. These include for example the creation of small company stock markets. The important contributions of stock markets for biotechnology companies are the liquidity provision to finance their activities and the opportunity they represent for venture capitalists to obtain their return on investment.

The Role of Foresight in the Governance of Biotechnology[5]

Foresight first rose to prominence in the 1980s when it was used by policy-makers primarily to predict potentially successful trajectories for science, technology and innovation and hence to improve national, sectoral or company competitive advantage. Its early goals were thus to inform decisions on the balance and direction of publicly funded science and technology, for example through strategic assessment of the opportunities and likely competitive challenges in particular key fields. The focus was upon tracking medium and long-term scientific and technological developments and their implications for research funding and the development and commercial exploitation of technological capabilities (Martin, 1995; Georghiou, 1996).

Subsequent changes in Foresight aims were partly driven by the failure of some science and technology policies that had neglected the socio-economic dimension, and were stimulated also by changes in the character of complex modern technologies. The salience of socio-economic issues reflects the greater complexity of the process of innovation and its organisational setting. Thus innovation is often dispersed across a network of knowledge and economic actors, encompassing a greater diversity of players with complementary knowledge and economic contributions, while the pervasive uptake and far-reaching social implications of innovation in life sciences gives the final consumer/citizen a potentially key role.

---

[5] This section is based in part on an EC-funded Accompanying Measure funded during the Fifth Framework Programme: Integrating Technological and Social Aspects of Foresight in Europe (ITSAFE), STPA-2001-00010. The Final Report and Annexes can be found on www.supra.ed.ac.uk/reports.

In recent years, Foresight has thus become bound up with shaping social, environmental and technological futures. There have been attempts to include the social forces shaping and shaped by scientific and technological developments. We have also seen the emergence of "social foresight" that takes as its starting point a set of societal problems (such as ageing and demographic change) rather than technological developments. For example, the British Government in October 1999 announced the launching of a new Foresight exercise completely reoriented towards "social demand" with emphasis on interactions within society and on assuring the value of science for society (UK Office of Science and Technology (OST), 1998).

Beyond engagement between science, technology and the social sciences, Foresight has further evolved into a process of charting societally desired future scenarios and exploring the social and policy processes that will foster their emergence, sometimes referred to as "social Foresight". More recent Foresight studies rarely seek to "predict" the future but rather to offer meaningful, plausible and pertinent visions of the future to permit and encourage societal actors (including scientific and technical experts, investors, decision-makers and citizens) to participate in the development of the future in a dynamic context of policy exploration, implementation and review.

A key issue for the development of life science policy is related to changing science and technology regimes (Grupp, 1993) where the borders between science disciplines are fading and technological advances are emerging at the interfaces between technologies. There is thus a need to rethink the role of R&D policy, no longer just to support particular technology areas but to adopt the role of moderator and to bring together key actors and technologies that may have a role in contributing to a particular desired future. The potential role of policy as an active moderator in the research arena requires not only top-down policy design, but also bottom-up approaches involving active moderation to facilitate co-ordination and develop guiding principles (Austrian Institute for Technology Assessment (ITA), 1998).

Analysis of governance and policy-related recommendations of Foresight studies indicates that there are pronounced differences between the healthcare and agro-food sectors in the life sciences. In the agro-food sector there is a strong emphasis on public acceptance, public information, safety and regulations. In healthcare such issues are not prominent among Foresight recommendations which tend to focus on how some areas, such as children's health or the relationship between environment and health could be improved. Likewise, public participation is not a big issue in healthcare Foresight although public involvement in discussions on healthcare technologies is taking place in different contexts. The UK Foresight Programme (Georghiou, 1996) is one example where lay voices are included in Foresight discussions on health-related issues. This seems to indicate that most Foresight studies perceive the healthcare system as a self-contained system, developing solutions for patients who do not interact with these developments. In agro-food related Foresight on the other hand, there are calls for stronger public participation in innovation processes, with implications for the integration of socio-economic issues into Foresight. This difference is likely to be related to the very different levels of public controversy in the two areas.

Integration Strategies in Foresight

The most frequent mechanism observed in the ITSAFE Project to integrate socio-economic and scientific-technical aspects of Foresight was via expert selection. Most studies included socio-economic experts in some or all of the Foresight process. Socio-economic experts were often integrated into all stages of the Foresight process including the definition phase where the issues were elaborated, the process of carrying out the exercise, and the interpretation and dissemination stage where the results of the process were evaluated. However, Foresight projects in the life sciences domain still seemed to be dominated by scientific and technological experts. Only a few studies explicitly ensured that socio-economic experts participated in carrying out the Foresight process per se. It was difficult in practice to evaluate the extent to which integration played a role in the assessment phase of Foresight projects, as this phase was often not clearly defined. Rather there were many follow-up Foresight activities where socio-economic expertise was involved, for example in workshops, conferences and seminars.

Another way to integrate socio-economic and scientific or technological dimensions in Foresight was to create a formal link, establishing a co-operative structure between organisations with different expertise. For example, an Austrian Foresight project was carried out as a co-operation between the Austrian Institute for Trend Analysis and Crisis Research (ITK) and the Institute for Technology Assessment (ITA) and the interaction between these organisations was the main channel for integrating socio-economic and scientific-technological perspectives (ITA, 1998).

Two aspects of the integration process can be related to sector specificities: the extent to which the definition of socio-economic issues depends on particular characteristics of different technologies; and the identification of sector-specific factors that impeded or are supportive of integration. In the ITSAFE project, we found a clear relationship between technology characteristics and the definition of socio-economic issues. Far-reaching, pervasive technologies such as genetic engineering can lead to the definition of a broad spectrum of relevant socio-economic issues (Georghiou, 1996; ITA, 1998; Jaeckel et al., 1995), including ethical, legal, environmental, economic and acceptance aspects. More narrowly defined technologies, on the other hand, lead to a more distinct pattern of socio-economic issues, for example the use of new enzymes for industrial processes raised almost exclusively economic issues such as possible price reductions, market shares or comparative positions (Menrad et al., 1999).

It is important to distinguish between (i) the integration of socio-economic and scientific/technical integration in Foresight processes and (ii) the involvement of stakeholders in Foresight. Three levels of engagement of stakeholders in the Foresight process in life sciences can be distinguished:

1. Discussion of the outcomes of Foresight activities with different stakeholders.
2. The involvement of stakeholders in the definition of Foresight issues.
3. The involvement of stakeholders in the entire Foresight process.

Level 1 involvement offers stakeholders no influence on the definition of Foresight issues and they do not have the opportunity to shape the process. This approach was commonly used in early projects which merely confronted stakeholders with the outcomes of Foresight.

In level 2 involvement, stakeholders participate in the elaboration of issues at the beginning of the Foresight project. This crucial stage shapes the direction of the process, and involvement of stakeholders allows the introduction of different interests and values into the problem-definition phase. However, the Foresight process itself is still mainly an expert activity where stakeholders' interests and values have no subsequent influence on the process.

Level 3, the highest degree of stakeholder involvement, allows stakeholders not only to participate in defining the problem and discussing outcomes, but also to engage actively in the ongoing process. Stakeholders can shape the outcomes of the Foresight activity and this is the most advanced of the emerging new paradigms where Foresight is used in the sense of outlook. This is no longer merely equivalent to prediction or forecasting of future developments, but takes into account that there is no single future and that different future options will be determined at least partly by different current policy choices. Foresight can then be seen as a means to inform and support such choices through its communicative and procedural character (Martin, 1995). This approach to Foresight has recently been elaborated (Grupp and Linstone, 1999) so that its elements are perceived as a means of communication for societal negotiation (Martin and Johnston, 1999).

This communication role for Foresight can be important where conflicting interests or values are involved in science and technology related issues. For example an Agro-Food Delphi study (Menrad et al., 1999) revealed pronounced differences of positions between stakeholders in assessing the application of GMOs in food production. While experts and stakeholders from science and industry made rather positive assessments, NGOs, consumers and farmers expressed more sceptical attitudes. On the other hand, in the Austrian Foresight process (ITA, 1998), conflicting interests did not show up to such an extent, supporting the notion of the influence of cultural and/or historical factors on Foresight processes. This result is particularly interesting given that, at the policy and political level, Austria is among the group of EU countries that is most opposed to GM crop technology.

Elaborating the conflicting views of stakeholder groups adds an important dimension to Foresight, generating knowledge on possible future conflicts associated within scientific or technological developments. This in turn can allow policy-makers to elaborate strategies in advance to deal with these likely emerging issues. In other words, Foresight can function as an early warning system enabling the development of policies in a precautionary way.

Compared to the stated need for integrative approaches to Foresight in the life sciences, and the responsibility of the public domain for initiating such activities, the ITSAFE analysis indicated that integration of technological and social aspects of Foresight in the life sciences was still at an early stage. Accordingly, current awareness of the significance of social/technical integration among key actors in life science related Foresight was not very strong. A few studies followed the deterministic approach while more recent studies exhibited instrumental coupling implying that the preconditions for better integration with a stronger socio-economic focus are at least favourable.

Conclusions

The long timescales involved in research and development in the life sciences create problems for companies to predict markets and technology trajectories successfully and to finance development over such long periods. The policy environment within which this technology is being developed therefore has a more than usually important impact on the sector's performance. Foresight (whether conducted by governments on behalf of an industry sector or for companies, usually multinationals, on their own account) has a useful role to play in helping to guide the strategies of industry and policy-makers over these long timescales. This is particularly the case given the turbulence of the operating environment for life sciences including increasingly fierce competition among companies on a global basis, and rapidly evolving stakeholder perspectives and interactions.

The results of the EPOHITE project described here suggest that variation in local circumstances, particularly in various policy instruments, targeted and generic, make a difference to the performance of the life science sector. However, another feature of the long development timescales in life sciences is that evaluation of the impact of policies on industry performance is also even more difficult than usual.

The EPOHITE project results showed that European member states can be grouped into four different clusters on the basis of biotechnology performance: (1) Denmark, Sweden and Finland, best-performing; (2) the Netherlands, United Kingdom, Belgium and Germany, second-best-performing; (3) Austria, France and Ireland, below the European median performance; and (4) Italy, Spain, Portugal and Greece performing weakly on all indicators.

Comparison of policies within and between these clusters shows the following:

- National policies for the biotechnology base and its commercialisation have a profound effect, either positive or negative.
- Policies to create and sustain the knowledge base are crucial for commercialisation but the reverse is not true.
- Countries with a systems perspective and a broad set of policies to promote biotechnology, addressing all the functions of the innovation system and

creating an environment conducive to entrepreneurial activity, achieve better performances than countries with patchy and fragmented policies.

- *Ex ante* co-ordination amongst strategic policy decision-makers (public or private) responsible for all the different functions of the innovation system can be extremely beneficial to developments at a national level and in avoiding policy gaps or overlaps.

Thus, policies to promote the commercialisation of biotechnology should not focus entirely on support for commercialisation. It is better to implement an integrated and balanced mix of instruments that target the creation and sustaining of a competitive biotechnology knowledge base and commercialisation. Moreover, policy-makers should monitor the future development of biotechnology in European countries. In some countries the first indications of a drying up of the knowledge base are already appearing and policy efforts to support a renewal of the knowledge base are needed. Designing biotechnology policies on an integrated basis implies that policy-makers should be aware of the strong and weak points of their national system and consider the (positive or negative) interactions of all policy measures that have an influence on the development of the innovation system.

Biotechnology is still at a stage where there are numerous possibilities for its future exploitation and diversity of competencies and policies is advantageous for driving future developments. Is there enough diversity among European countries' biotechnology competencies and what role should European policy play in relation to that of member states' policies? This is an area where further research is needed on the role of diversity in specialisation in scientific and commercialisation patterns within Europe for the future competitiveness of European biotechnology.

Following a period where countries focused on encouragement of start-ups, a selection phase is now likely where only a limited number of firms will be able to move towards sustainable growth. Should this be an independent market-driven process or should policy intervene? Direct support for the growth of firms would interfere with market mechanisms. However, the creation of favourable framework conditions for growing biotechnology firms is desirable, including support for the recruitment of scientists, regulation in general and IPR issues in particular, consideration of ethical and consumer issues, and conditions for private financing.

Foresight can play an important role in encouraging networking and the formation of new partnerships among various stakeholders, in particular from science and industry (e.g. Georghiou, 1996; Menrad et al., 1999). Sometimes regional tensions and national divergences can lead to engagement in economic and technological Foresight in the service of strategic political and economic objectives. Foresight practices can thus reflect the tensions within the societies whose potential they seek to portray and can enable scientists, companies and policy-makers to devise more effective and more integrated policies.

Treating Foresight as a process of extrapolation from social requirements does, however, suffer from some intrinsic problems:

- We often have only a partial prior understanding of the socio-economic implications of a new technology.
- Current public preoccupations tend to be projected onto emerging technologies so that the technology operates as a kind of Rorschach Test for social concerns.
- Some criteria tend to be given more weight than others, in particular risk avoidance.
- Issues of trust have become central in the context of life sciences, where the proponents of a technology are seen as having provided unduly favourable assessments of the outputs from an industry sector.

Two types of relationship between Foresight and policy activity can be differentiated: a *direct* relationship in the sense that recommendations are taken up by policy-makers and transformed into activities, such as reorientations of research funding or technology support programmes; and an *indirect* relationship where the outcomes of a Foresight exercise make it possible for policy-makers to consider new options or open up new approaches and arenas which are guided only indirectly or partially by the original Foresight process.

Foresight in the public domain is generally directed towards informing public policies and in some countries it has become part of national moves towards more integrated approaches to policy-making across a broad range of issues (Lyall and Tait, 2004). To some extent, this is dictating the pace of evolution of Foresight, attempting to deliver adaptive policies that can respond rapidly to emerging developments. The integration of social and technological aspects of Foresight suggests that we will need to combine different kinds of evidence and argument in new, more creative ways.

One consequence of this is that policy needs to be an active moderator, bringing together key actors and technologies, and leading to an increasing focus on networking between research organisations and industry, financial services and other actors and particularly involving societal stakeholders. However Foresight, even if well integrated, should not be expected to resolve conflicts that are already well developed and highly polarised. Indeed it can further increase the degree of polarisation.

The key to effective use of Foresight in developing policy and guiding science and industry strategies is to conduct the integrated Foresight exercise before full-blown controversy has emerged and to have key actors, including policy-makers, scientists and industrialists committed to implementing its recommendations.

References

Allansdottir, A., Bonaccorsi, A., Gambardella, A., Mariani, M., Orsenigo, L., Pammolli, F. and Riccaboni, M. (2002), 'Innovation and Competitiveness in the European Biotechnology', Enterprise Papers – No 7, report commissioned by the European Commission, DG Enterprise, as background paper for the Competitiveness Report 2001.

Carlsson, B. and Jacobsson, S. (1997), 'In search of useful public policies: key lessons and issues for policymakers', in Carlsson, B. (ed.), *Technological Systems and Industrial Dynamics*, Dordrecht: Kluwer Academic Publishers, pp.299-316.

Dominguez Lacasa, I., Reiss, T. and Senker, J. (2004), 'Trends and gaps of biotechnology policies in European Member States since 1994', *Science and Public Policy*, in press.

Enzing, C.B., Benedictus, J.N., Engelen-Smeets, E., Senker, J., Martin, P., Reiss, T., Schmidt, H., Assouline, G., Joly, P.B. and Nesta, L. (1999), 'Inventory of public biotechnology R&D programmes in Europe', Vol. 1, Analytical report. Luxembourg: Office for Official Publications of the European Commission.

Enzing, C.M., van der Geissen, A.M. and Kern, S.J. (2002), *Life Sciences in Nederland*: Economische Betekenis, Technologische Trends en Scenario's voor de Toekomst, STB-02-23.

Enzing, C.M., van der Geissen, A. and Kern, S. (2004), 'Commercialisation of Biotechnology: Do dedicated public policies matter?', *Science and Public Policy*.

Ernst and Young (2001), 'Integration', Eighth Annual European Life Sciences Report, London.

European Commission (1996), Green paper on innovation, Luxembourg.

European Commission (2001), European competitiveness report, Luxembourg: Eur-Op, 2001, 142p., catalogue no. NB-39-01-110-EN-CEN.

European Commission (2002), 'Life sciences and biotechnology – a strategy for Europe'.

European Commission (2003), 'Biotechnology Innovation Scoreboard 2003', European Trend Chart on Innovation, EC Enterprise Directorate-General, Brussels: ftp://ftpnl.cordis.lu/pub/trendchart/reports/documents/report7.pdf.

Georghiou, L. (1996), 'The UK Technology Foresight Programme', *Futures* **28**, No 4.

Grupp, H. (1993), 'Technologie am Beginn des 21 Jahrhunderts', Heidelberg: Grupp, H. (Technik, Wirtschaft und Politik), p.265.

Grupp, H. and Linstone, H.A. (1999), 'National technology foresight activities around the globe: Resurrection and new paradigms', *Technology Forecasting and Social Change* **60**, pp.85-96.

Hermans, R. and Luukkonen, T. (2002), 'Findings of the ETLA Survey on Finnish Biotechnology Firms', Keskusteluaiheita – Discussion papers, No. 819, 39 pp.

Howell, M., Trull, M. and Dibner, M.D. (2003), 'The rise of the European venture capital for biotechnology', *Nature Biotechnology* **21**(11), pp.1287-1291.

ITA (1998), Delphi Report Austria 1: Technology Delphi I – Konzept und Uberblick. Vienna: Institut fur Technikfolgen-Abschatzung, p.126.

Jaeckel, G., Menrad, K., Reiss, T. et al. (1995), *Die Zukunft des Deutschen Gesundheitswesen aus der Sicht von Artzen und Experten*, Karlsruhe: Fraunhofer ISI, p.422.

L'Heureux, K., Libeau-Dulos, M., Nilsagard, H., Rodriguez Cerezo, E., Menrad, K., Menrad, M. and Vorgrimmler, D. (2003), 'Review of GMOs under Research and Development and in the Pipeline in Europe', Karlsruhe: Seville.

Lyall, C. and Tait, J. (2004), 'Foresight in a Multi-level Governance Structure: policy integration and communication', *Science and Public Policy*, **31**(1), 27-37.

Mangematin, V., Lemarie, S., Boissin, J.P., Catherine, D., Corolleur, F., Coronini, R. and Trommetter, M. (2003), 'Sectoral system of innovation, SMEs development and heterogeneity of trajectories', *Research Policy*, **32**(4), pp.621-638.

Martin, B.R. (1995), 'Foresight in Science and Technology', *Technology Analysis and Strategic Management*, **7**, pp.139-168.

McKelvey, M., Alm, H. and Riccaboni, M. (2003), 'Does co-location matter for formal knowledge collaboration in the Swedish biotechnology-pharmaceutical sector?', *Research Policy*, **32**, pp.483-501.

Menrad, K., Agrofiotis, D., Enzing, C.M. et al. (1999), 'Future Impacts of Biotechnology on Agriculture, Food Production and Food Processing', *A Delphi Survey*, Heidelberg: Physica-Verlag.

OST (1998), 'Blueprint for the next round of Foresight', *DTI*, December, DTI/Pub 3733/65k/12/98/NP. URN 98/1032.

Reiss, T., Calvert, J., Dominguez Lacasa, I. et al. (2003), 'Efficiency of innovation policies in high technology sectors in Europe' (EPOHITE), Luxembourg: Office for Official Publications of the European Communities.

Sandstrom, A. and Norgren, L. (2003), 'Swedish Biotechnology', *scientific publications, patenting and industrial development*, VINNOVA Analysis VA 2003:2.

Scottish Enterprise (2001), 'Biotech Scotland – Framework for Action 2001-2002', Glasgow: Scottish Enterprise.

Senker, J., van Zwanenberg, P., Enzing, C., Kern, S., Mangematin, V., Martinsen, R., Munoz, E., Diaz, V., O'Hara, S., Burke, K., Reiss, T. and Worner, S. (2001), 'European Biotechnology Innovation System', Brussels: EC.

Storey, D.J. and Tether, B.S. (1998), 'New technology-based firms in the European Union: an introduction', *Research Policy*, **26**(9), pp.933-946.

Tait, J. and Williams, R. (1999), 'Policy Approaches to Research and Development: Foresight, Framework and Competitiveness', *Science and Public Policy*, **26**(2), pp.101-112.

Tait, J., Chataway, J. and Wield, D. (2004), 'Governance Policy and Industry Strategies: Agro-biotechnology and Pharmaceuticals', Innogen Centre Working Paper 12, July, http://www.innogen.ac.uk/public.php.

Ward, M. (2001), 'UK vs Germany, Playing catch-up, softly', *BioCentury*, pp.A1-A4.

Chapter 5

Developing an Integrated Approach to Risk: The ILGRA Network

James McQuaid

Risk is a prime example of an issue that pervades policies across government and for which there is a recognised need for integration of policy-making. There is an inherent difficulty in seeking to achieve integration by the pursuit of evidence-based policies especially on the increasingly contentious risk issues arising from the collision of scientific developments and public unease about them. The difficulty derives from the fact that risk is itself a representation of uncertainty about causes and consequences of activity. The evidence, such as is usually available in the form of expert assessment rather than scientifically established facts, will often be contested as to its reliability and relevance. This contest will be characterised by varying and conflicting judgements about the significance of a particular risk and hence about the consistency of handling policies compared to those adopted in other contexts. This lack of objectivity of the evidence has to be countered by dialogue and deliberation involving actors and stakeholders inside and outside government. The organisation of this dialogue and the mechanisms needed to ensure that information is fully shared and that the dialogue is inclusive of the interests involved are important aspects of the governance of risk. A necessary condition for effectiveness is discipline and order in the application of procedures and practices for assessment of risk and of the costs and benefits of measures to control risk. Furthermore, the criteria and value judgements that influence the evaluation of risk must inevitably figure in the process of dialogue. All of these considerations dictate a need for transparency and openness if consistent policy outcomes are to be achieved.

Risk issues do not as a rule map uniquely on to the policy remits of individual government departments and this gives rise to the possibility of lack of coherence and consistency of decisions in different policy contexts. The recognition of this motivated the setting up in 1992 of a network of officials involved in risk policies in UK government departments and agencies. Over a period of several years, the network strengthened its working methods and adopted a diversity of approaches in support of the process of informed dialogue and interaction inside and outside the network. It retained its informal status, the informality being related to it not having an official locus in the machinery of government and consequently it did not have a formally agreed programme of work. Despite this, it succeeded in gaining recognition and in anchoring its agenda to various high-level policy imperatives relating to risk. Its achievements in various aspects of policy integration on risk issues have been influential and widely acknowledged. The question of how all this was brought about provides an interesting case study on policy integration and forms the subject of this chapter. Particular regard is paid to the status of the network as an instrument of governance rather more than as developer and proponent of ideological positions on how risks should be regulated.

The ILGRA Network

The Interdepartmental Liaison Group on Risk Assessment (ILGRA) was established in 1991 on the initiative of the Health and Safety Executive, the regulatory body for most industrial risks in the UK. The initial purpose was to bring together departments and agencies for the practical task of co-ordinating government input to an international conference on risk assessment held in London. The evident benefits from the initiative resulted in a decision in 1992 to keep the group in being as an informal network to maintain the liaison role.

Kinds of Network

The ILGRA network in liaison mode was characteristically weak in terms of the interdependencies between the participants. Such weak networks exist mainly to enable interactions between participants and to improve communications and information exchange whilst the participants independently pursue their own objectives. The output of the network is the simple summation of the contributions that participants produce in their respective spheres. The contributions are brought together by a co-ordinator or exchanged through meetings and taken up, or not, by the other participants as they wish. The benefits of the network come from breaking down barriers and improving access to information in the hope of reducing duplication of effort and improving learning from the experience of others. The process at work is, at best, passive adaptation depending on whether a feature that matters, and for which solutions have been found in one context, is perceived to matter in another, with no traceability of the reasons for adaptation or not. Many research networks are of this kind as of course are those in the normal run of government co-ordinating committees. They achieve little in terms of active integration and joined-up working.

For the latter to happen, a network needs strengthening by linking of resources and by defining relationships that create interdependencies between participants and between activities. This kind of network of interdependency then has the potential to produce more than the summation of individual contributions produced in isolation from each other. For example, the integration of resources may enable a "critical mass" of resources to be deployed on a priority task that any one participant would not be able to undertake on their own. Furthermore, it is the nature of the policy integration problem that there exist issues contiguous to the main policy applications and which may not be of high priority in any one context but nonetheless are of common occurrence across the domain of the policy driver, in this case risk. When the implications are integrated across the domain, such an issue may assume a generic importance that warrants collective action. An example, to be discussed later, is that all risk regulators have good reason to understand better why some risks capture public and media attention and others do not, but any one regulator is not sufficiently motivated to investigate such a generic issue. Passive adaptation cannot apply since there is no one doing anything from which others can learn. A further benefit of the stronger kind of network is that the relationships forged between participants can be useful in promoting interdisciplinary working. This is a very desirable result given that risk issues invariably have a number of dimensions that are vulnerable to capture by professional disciplines with consequent restrictions on the perspectives applied to the problem.

A relevant comparison of the two kinds of network from a quite different area is between the strong networks being fostered under the European Commission's 6th Framework R&TD programme as "networks of excellence" and the weak networks that characterised the "shared cost" actions of earlier Framework programmes.

Transformation of ILGRA

The trigger for the transformation of the ILGRA network from liaison to interdependency came fortuitously. It arose from a commitment by the Government to undertake a review of the principles and practices used in government for risk assessment with a view to identifying best practice, encouraging common approaches and to producing a report describing and comparing current practices. The commitment had its origins in the Government's strategy for the environment published in the 1990 White Paper "This Common Inheritance" (HMSO, 1990). The strategy identified certain risk-related principles relevant to the present discussion, in particular that:

- decisions should be based on the best possible scientific information and analysis of risks, and
- where there is uncertainty and potentially serious risks exist, precautionary action may be necessary.

The strategy recognised in relation to practical implementation that, in policy-making, proper tools of analysis have to be applied but importantly it also acknowledged that judgements have to be made about the weight to be put on the influencing factors in particular cases.

The Government's commitment to undertake the review was contained in the second annual progress report on the strategy (HMSO, 1992). The existence of the ILGRA network and its desire for an activity on which to anchor its collective expertise led to the acceptance of its proposal that it should have responsibility for conducting the review and preparing the report for ministers. The group thereafter took a broader remit embracing all aspects of risk pertaining to policy-making including the development of advice on technical and policy aspects of risk assessment, evaluation, management, perception and communication. However, it retained its original restricted title, a source of subsequent confusion in respect of both the terms liaison and risk assessment. Rather unconventionally, it continued to operate as an informal group without official status despite reporting to and receiving endorsement of its proposals from ministers.

The risks covered by the review were not restricted to those relevant to the environment. The need for a wider-ranging review was a recognition that sources of risk do not have unique and separate effects confined to a single target such as the environment. For example, a pesticide has the potential also to affect workers and the public through various pathways yet responsibility for regulating risks has to accommodate to the boundaries of departments, each with a large degree of autonomy. In such circumstances, there is always a danger that departments or regulatory agencies may adopt too narrow an outlook in regulating the risks within their area of responsibility. There are obvious practical limits to achieving policy integration through the centralising of authority in single departments or agencies, though steps in that direction have been taken, for example the HSE for nearly all work-related risks and the Food Standards Agency for all food risks. An alternative is to seek integration through the adoption of a common regulatory philosophy or framework founded, for example, on an overarching principle such as "As Low As Reasonably Practicable" (ALARP) allied with an emphasis on responsibility for control being placed on those who create the risks. However, the approaches to managing and regulating risks had not been developed systematically in government but had evolved over time within departments. The situation in 1992 was thus that the position reached in this evolution was ripe for a review of the extent to which policy integration, in the sense of operating to a common framework of methods and practices, had been achieved in the circumstances of the institutional arrangements that prevailed. A detailed study of the evolution of these institutional arrangements was carried out at a later date by Hood et al. (2001).

From that beginning, ILGRA developed to become an effective network, facilitating improved policy integration through a variety of mechanisms. The nature of the network changed with time and it spawned some satellite networks dealing with specific issues. It was instrumental in commissioning several multi-departmental research projects addressing issues relevant to a common framework of regulation and in improving links with external bodies through open seminars and lectures. It developed linkages with various other government strategies, for example on science and technology and on better regulation. The work of ILGRA laid the foundations for the report of the Cabinet Office Strategy Unit of 2002 on Risk (Strategy Unit, 2002) which might be regarded as providing the guidelines on policy integration called for in Chapter 1 of this book.

The description that follows begins with a view of the characteristics of ILGRA as a network and follows with illustrations of some of the mechanisms it adopted to fulfil the aim of encouraging policy integration. The description relates to the activities up to the time when the author ceased to be chairman of ILGRA at the end of 1999. A discussion is included on the impact of ILGRA's activities together with some remarks about recently implemented changes in responsibility for policy integration on risk. A related analysis of the factors affecting the development of integrated policies on risks has been given elsewhere together with further background on the role of ILGRA (McQuaid and Le Guen, 1998). Interesting and relevant insight into the challenges of risk regulation from a ministerial perspective has been given by Waldegrave (1987).

Characteristics of the ILGRA Network

Although much emphasis is placed in network theory on the forms and structures of networks, this aspect is not considered to be of any great relevance to the purpose of this chapter. Rather, it suffices to note the view of Gillies and McCarthy (2002) that "a network is as much about a way of behaving and being as about a particular organisational form". Taking this as a cue, the characteristics of the ILGRA network that appear most relevant to illuminating its ethos may be listed as follows:

- among its participants, it possessed a wide diversity of professional disciplines (scientists, engineers, economists, sociologists ...), occupational roles (policy generalists, scientific and economic advisers, inspectors ...) and, of course, departmental interests (almost all departments and agencies from central government and devolved administrations);
- there was mutual trust and absence of competition amongst participants who were all equally valued;
- there was a considerable degree of continuity in the membership but sufficient inflow of fresh minds to challenge any tendency to group thinking;
- deliberation on issues was marked by candour and willingness to expose weaknesses in positions;
- meetings were arranged at a neutral venue to avoid any impression of capture by a host department;
- although the HSE provided the chairman and secretariat, this reflected the need for the network to have an agent to organise proceedings and joint activities and to fulfil certain tasks such as production and publishing of reports;
- the network was not homogeneous in terms of participant engagement; some participants were strongly connected and others less so, reflecting the strength of their departmental interest;
- participants did not merely operate to departmental briefs but contributed according to their expertise, drawing as necessary on their departmental backgrounds. It was up to them how and to what extent they consulted in their departments, in keeping with the informal status of the network.

It will be evident from the above listing that the characteristics of the ILGRA network owed much to the characteristics of the individuals who comprised it. Many were of long experience in policy-making on risks and of operating across departmental boundaries and with external interests, particularly in international forums. In addition, the influence of HSE participation may have been subliminally important given that this nominally-integrated organisation was still coming to terms with the fragmented responsibilities for risk regulation in different industrial sectors that it had inherited on its establishment nearly 20 years before.

Use of Risk Assessment in Government Departments

The event that largely set the network's agenda was the conduct of the review described earlier. The review was in two stages: a fact-finding exercise on practices in departments and a follow-up exercise probing what departments felt was necessary for developing a common framework of risk policy-making and exploring areas of common ground in more detail. The results have been described in detail in ILGRA (1996). The review showed that, although there was a considerable commonality of approaches, there were areas where greater coherence and consistency would lead to tangible benefits. The main areas identified for action were:

- the adoption of more consistent methodologies in order to avoid conflicting approaches to assessment of similar risks and to provide a robust framework for decisions on the tolerability of a risk;
- the adoption of a common terminology, insofar as this was feasible given the rigidity of established usages in different areas of risk;
- the examination of common approaches to cost-benefit analysis;
- the institution of arrangements for the exchange of data and information about risks to ensure that departments operated with the same information base when assessing similar risks;
- a universal acceptance of the need to improve the management of risk communication with the public, with the emphasis on a two-way process of engagement and debate supported by a policy of transparency and explanation in place of one-way provision of information;
- the need for methods of ranking risks to assist the setting of priorities, taking account of the perceptions of the public;
- the need for greater co-operation on research in place of, or supplementing, compartmentalised programmes within departments on topics of wider interest;
- the development of a common UK approach to risk assessment at an international level to improve the UK's effectiveness in contributing to the development of standards.

This extensive agenda was endorsed by ministers with a request that progress should be reported at two-year intervals. The Government in its 1995 "Forward Look of Government Funded Science, Engineering and Technology" (HMSO, 1995) signalled this endorsement and, specifically, that ILGRA would be asked to put forward proposals on how the issues should be addressed.

Delivery of the Ministerial Agenda

There was no fixed timetable or order of priorities for advancing the various items agreed by ministers. In practice, there was some interdependence of items; for example, a research project might need to be organised and delivered before a recommended methodology or practice could be prepared.

An account of progress was given in a report to ministers in 1998 (ILGRA, 1998). The report also included an analysis of changes in the policy background that had taken place, such as the evidence of a growing lack of trust by the public regarding some scientific and technological developments and the vigorous debate taking place on the need for balance between government intervention on risk issues and the freedom of choice of individuals. The implications of the changes for departments and the consequent need for expansion of ILGRA's agenda were drawn out.

The report on progress concentrated on four of the broad areas on the agenda:

- the development of more consistent methodologies;
- improving communication;
- sharing knowledge and information, and
- commissioning multi-departmental research programmes.

Detailed accounts are given in the ILGRA report and related reports cited therein and will not be repeated here. The matters of particular interest to this discussion are the ways in which the work was progressed as illustrations of joined-up working. These mainly related to the development of more consistent methodologies on which the discussion below will concentrate. The activities under this heading employed two subgroups or satellite networks of the main ILGRA network and two studies commissioned from external organisations.

Risk Assessment and Toxicology

The first of the subgroups was given a remit that linked ILGRA's programme to the 1993 White Paper on Science, Engineering and Technology (SET). A commitment was given in the White Paper that an annual Forward Look on Government funded SET would be published and the first of these (HMSO, 1995) described a long-term exercise to be undertaken under the aegis of ILGRA on risk assessment and toxicology. The task was to carry out a mapping of the current practices for managing risks to health from toxic substances and to follow this with a long-term research strategy to develop and validate improved methodologies taking advantage of recent advances in scientific knowledge and techniques. The objective was to reduce the usage of laboratory animal testing in toxicological assessment. The ILGRA subgroup, under the title Risk Assessment and Toxicology Steering Committee, brought together the departments with a policy interest of which there were many; exposure to chemicals impacts on the regulation of food safety, manufacturing processes in industry, agriculture, the safety of medicines and of products used in the home and workplace, and environmental pollution. The subgroup also included members from the relevant Research Councils. It was led by the (then) Ministry of Agriculture, Fisheries and Food and it engaged with external experts in several workshops addressing particular aspects of the task. The organisation and reporting of the work of the subgroup (see, for example, IEH, 1999) was contracted to an independent academic institute. Although the subgroup operated under the ILGRA umbrella, it was funded directly by contributions from the departments involved.

Methodologies for Setting Safety Standards

The second subgroup focused on methodologies for the setting of safety standards, a topic at the heart of policy integration on risk. The task of the subgroup was to follow up the recommendations of a study on the setting of safety standards carried out by an interdepartmental group led by HM Treasury and supported by a technical group of external advisers (HM Treasury, 1996). The Treasury group had originally been established independently of ILGRA but it was agreed that it would join with ILGRA to progress its recommendations though it would continue to be led by HM Treasury. The Treasury study had its origins in the need to examine a widely-held belief that safety standards ought to be based on the utilitarian view that a single figure for the valuation of a statistical life (VOSL) should be applied uniformly across all areas of safety regulation. The idea had been imported from the US where a study at the time had shown an enormous range of values imputed by US regulations. The Treasury study concluded that a fully rule-based approach to safety regulation, where all regulations would be set according to universal formulae quantifying costs and benefits, would be unrealistic. Nonetheless, the study concluded that common frameworks, including other considerations besides utility, could and should be developed to provide a common basis for policy judgement and standard setting.

The subgroup, which also included external advisers, set out to evaluate institutional procedures and conventions by which departments set safety standards. They examined three case studies related to environmental regulation and included in the evaluation the extent to which the setting of safety standards adhered to the guidelines of the Cabinet Office's Better Regulation Unit. Two of their conclusions were of particular importance in relation to later developments. The first concerned the role of scientific advisory committees and the lack of transparency on how they reach decisions about levels of risk. It was felt particularly important that the terms of reference of scientific advisory committees should make clear that their role is to assist decision-makers and not to set standards as such. This would ensure and make transparent that other considerations, such as costs and benefits, are also taken into account in the final decision. The second conclusion concerned the need to ensure that the standards adopted reflect the values of society at large. In addition to information on technical factors, policy-makers must give due weight to public perception of risk. Departments needed to develop ways of explaining to the public the policy as well as the scientific background to decisions and to establish how best to incorporate both expert judgements and society's preferences into the decision-making process.

Monetary Values for Avoidance of Fatalities

This major research project on a controversial and methodologically difficult subject was referred to in the report, though the research at the time was still in the early stages. The purpose was to produce well-grounded information on people's willingness to pay for reducing risks. The research used focus groups to access public beliefs and attitudes on such issues as the acceptability of cost-risk trade offs, essential to the methodological questions of the figure for the VOSL and the universality or otherwise of the VOSL in different risk contexts. The research, as eventually reported (Chilton et al., 2000), was successful in producing consistent evidence by a number of independent procedures to support a figure for the VOSL for the benchmark case of avoidance of road accident fatalities. However, the research also found that there was no empirical justification for a premium on that value when it was applied to other risk situations such as the railways or domestic fires, though that conclusion has been the subject of some controversy about its relevance to policy decisions. The results of the research have been highly influential; they were, for example, of central importance in the debate about decisions on rail safety measures that took place at the Public Inquiry following the Southall and Ladbroke Grove rail accidents.

International Negotiations

The remaining issue of methodologies concerned the development of a common UK approach to negotiations on risk assessment at international level. The issue arises because most regulations on health, safety and environmental standards now originate in European directives. The subject was addressed in a study commissioned by HSE on behalf of ILGRA. The study (HSE, 1998) found that there were wide differences in analytical approach, between policy fields and between institutions, within UK government and within the EU. In particular, in developing European and UK regulations, the criteria for policy choice were often unclear. In view of the shortcomings found in the domestic UK approach, the study devoted most of its attention to the actions needed for improvement in the UK context. It suggested that only limited progress towards greater consistency could be expected from the efforts of individual departments in their own fields.

The study concluded that a strengthening of central, Whitehall-wide machinery was needed but ILGRA in its existing form had limited potential for achieving reform, lacking authority or sanctions. Nonetheless, it suggested that ILGRA could be more effective if it developed a stronger sense of joint ownership by departments and operated at a more senior level. The study placed particular emphasis on the need to consider what a common framework should mean in practice – the study report providing useful pointers in that direction – and how it should be progressed. The consideration of these recommendations was deferred pending the outcome of a recommendation to ministers (see below) that departments should prepare statements of the frameworks they use, as a necessary precursor to the development, by whatever means, of the desired common framework.

Updating the Agenda

The progress report drew attention to the wide acknowledgement that managing risks solely on the basis of a probabilistic estimate of physical harm was unlikely to succeed. Public values and perceptions of risk must also be integrated in the decision-making process, though how far these can be taken into account in any particular case would depend on constraints (such as the need to meet EU obligations) and in some instances their legitimacy if they conflicted with principles of fairness and equity. Furthermore, whatever the decision-making process adopted, there would still be a need for balance between the extent to which risks are prevented or controlled, the resources required and the economic benefits foregone. In keeping with the conclusions of the work that had already been undertaken, two priority strands were identified. The first was the development of structures or frameworks for integrating risk estimates, public perceptions, the need for trade offs and other factors in the decision-making process. The second was the steps that departments and regulators could take to gain acceptance of their frameworks and of the consequent decisions as representing the best trade offs in particular situations. A necessary preliminary was to assemble an inventory of the frameworks used by departments in making decisions on risk. This, in effect, changed the approach to developing a common framework. The approach being pursued had been one of analysis, prescription and advocacy by ILGRA, having identified the need in the review described earlier. It was now to be one of consultation with departments and integration of the results by the Cabinet Office Better Regulation Unit working with ILGRA. The ILGRA progress report indicated the procedures that the departmental frameworks should describe, including importantly the active engagement of stakeholders in all stages so that they can influence the assumptions and value judgements that permeate the whole procedure and hence concur more readily with decisions emanating from it.

These proposals were endorsed by ministers, as were some new actions for ILGRA signalled in the report. These latter were influenced by awareness that the departmental frameworks would be likely to highlight some differences of approach. They included:

- clarifying the role of expert advice in the decision-making process. For this purpose, an interdepartmental research project was being set up to identify and categorise current practices from examination of a number of case studies and to draw up principles of good practice for the engagement of experts, the elicitation of their advice and for the incorporation of the advice in decision making. The research, which had the agreement of the Chief Scientific Adviser, was to be led by HSE and funded by ILGRA members;
- developing a consistent policy on a precautionary approach. The need for this had been identified in the 1990 White Paper on the environment referred to earlier. The issue was the increasing reference to the Precautionary Principle in international treaties and conventions and the considerable confusion that existed on the form it should take in operational terms. The action for ILGRA was to examine, within its network, the place that a precautionary approach should take in the decision-making frameworks that departments were being asked to articulate;
- commissioning further collaborative research. This mechanism for supporting joined-up working had proven successful and a wide range of potential topics for further research was identified. Projects on two topics may be mentioned that were linked to the preparation of frameworks and were subsequently put in place. The first, in collaboration with the Economic and Social Research Council, was to examine why some risks capture the public imagination whereas others are largely ignored. This phenomenon, known as the social amplification or attenuation of risk, can have serious knock-on effects, for example loss of confidence in institutions and diversion of resources. The work was intended to provide information on the role of the media and other institutions in shaping risk perceptions. The practical application would be to help departments to structure their risk communication policies and practices to take into account possible amplification or attenuation effects. An important innovation in this project was the engagement of US experts in the research since the subject had been much studied in the US context. The second project was on the contribution to achievement of trust from active engagement with and participation of the public both in the elicitation of value preferences and in the consideration of policy options. Several protocols were coming into use, for example focus groups and citizens' juries. The research sought to evaluate the effectiveness of the different protocols in different risk settings, again to assist departments in preparing risk communication strategies.

The above description of activities completed or in progress is not a full account of the studies and projects undertaken. Nevertheless, the examples serve to illustrate the operating style that was adopted for gaining a better understanding of the complexities of government policy-making on risk issues. The first step was to analyse an issue or possible course of action in a rigorous fashion, preferably by examining a range of case histories to establish what works and what does not. This enables common principles and procedures of good practice to be identified. In describing the agenda, emphasis was naturally given to tangible output-related activities since these allow the performance of ILGRA to be more easily reported and assessed. However, tangible outputs were not the only consideration. The agenda also included explicit measures to achieve desirable outcomes in the form of improved awareness and learning processes within government and the fostering of two-way interaction between government and the external community of experts on risk both in the UK and abroad. The former processes included deliberation within the ILGRA network itself and preparation of discussion documents for distribution to departments. The intention was to seek to identify the interconnections and feedback loops that would make the approach of one department depend on those of other departments or the handling of one risk interact with the handling of another, all in order to reduce the possibility of unintended consequences. A particular example in this regard was the extended programme of work across departments and Research Councils on toxicological risk assessment described earlier. The external interactions involved open lectures and seminars with invited experts and workshops with open publication of proceedings. Thus, overall, the operational style adopted for ILGRA can be argued to have predated the recent growth of interest in the application of systems thinking in government (see, for example, Chapman, 2002).

Discussion

The chapter began by describing ILGRA in terms of a network. Although much has been written about networks as an organisational concept (see for example Gillies and McCarthy, 2002), there appears to be little in the way of objective methods of assessing their effectiveness as compared to other structures. Evaluation of the ILGRA network has therefore to be subjective and no doubt the author's assessment will be, or be seen to be, biased in its favour. However, this bias is conditioned by experience of many other networks, both in the UK and internationally, in which the author has been involved or has knowledge.

In considering the effectiveness of ILGRA, the key questions for discussion are: to what extent was ILGRA successful in the process of joining up policy rather than being just another government co-ordinating committee; what actual impact did it have in the period under review and where does ILGRA now stand in the scheme of things?

On the first of these questions, the success of ILGRA in influencing the process can be described under the following headings:

- Style: The particular operational style it developed, characterised by active participation as described above, was well suited to the complex nature of risk issues. The operational style motivated departments to be engaged and to contribute resources.
- Relevance: It set an agenda that resonated with ministers' wishes for joined up working and its achievements towards that end were periodically reviewed and accepted by ministers.

- Self-adaptation: Although the Government has several important roles in handling risk, as described in Strategy Unit (2002), there was at the time no centrally-placed sponsor for risk to whom the role adopted by ILGRA would naturally have fallen. The motivation for action came from the collective view of departments, brought together by HSE, that action was needed especially in relation to the regulatory role of government. It was thus an illustration of self-adaptation by actors in the system to a perceived need.
- Learning: ILGRA fostered a positive approach to learning. Its informal status removed inhibitions and countered defensiveness. It encouraged explanation, debate and identification of areas of ignorance. The learning experience was disseminated outside ILGRA's own boundaries. For example, ILGRA was instrumental in establishing a training course on risk communication via the Civil Service College. How far the learning permeated through departments is still an open question since it is by its nature a long-term process of gradual improvement and culture change.
- Innovation: ILGRA was particularly innovative in collaboration across departments on research. The concept of multi-departmental funding of research, as distinct from sponsorship via the Research Councils, was at the time unusual if not unique. It differed from "club" funding, where a consortium of interests would meet the costs of a project with objectives defined by a promoter who was invariably the contractor who would undertake the work. Research proposals emanating from ILGRA integrated the requirements of the customers and the work was commissioned by competitive tender, with active involvement of the customers at all stages.
- Transparency: ILGRA provided a gateway to government for external interests. Its existence and its work became widely known amongst the growing national and international community of experts on risk through its lectures and seminars. The structure and processes of government are often opaque to outsiders and the forums provided by ILGRA contributed to transparency. To an extent, ILGRA was better known outside government than within.

Turning to the question of the actual impact of its work, there are some areas where the answers are clear enough as of now. The research it facilitated on VOSL provided much needed underpinning of the government-wide benchmark (on the valuation of road safety improvements) as well as producing evidence of public attitudes to relative adjustments of the benchmark for other areas of risk. The research on social amplification of risk led to the drafting of new guidance on risk communication that was "well regarded by risk management professionals within government" (Strategy Unit, 2002). The identification of the need for clarification of the role of scientific advisory committees, also raised in the report of the BSE Inquiry (Phillips et al., 2000), was subsequently progressed by the Office of Science and Technology through the development of a Code of Practice. This drew in part on the research into good practice that had been funded by a consortium of ILGRA member departments. The proposal for guidance on the application of the Precautionary Principle came to fruition and informed the Strategy Unit's report on risk (Strategy Unit, 2002). Supporting evidence of the influence of ILGRA is provided by the report on Science and Society from the House of Lords Committee on Science and Technology which identified ILGRA as the body to conduct a review of research on how risk information is received by the public (House of Lords, 2000). The same report also quoted a favourable view of ILGRA given in evidence to the Committee by the Consumers' Association. However, it is still moot whether the connections made between departmental policies as a result of ILGRA's work have, in practice, resulted in the desired self adaptation within departments. A proper evaluation would have to examine on a case-by-case basis whether and how departmental policies have been adjusted. It is probably too soon to expect to be able to do much of that and, in any case, recent changes in the scene, to be discussed below, will make it increasingly difficult to make a causal connection with the activities of ILGRA.

The event with the greatest potential impact was the gaining in 1998 of ministerial endorsement of ILGRA's suggestion that frameworks for the management of risks should be developed and published by individual departments. The Government Response to the Report of the BSE Inquiry (DEFRA, 2001) noted that most of the frameworks had by then been published and that:

- ILGRA and the Cabinet Office had conducted an initial survey of the risk frameworks to assess robustness and fitness for purpose;
- ILGRA, in conjunction with the Cabinet Office and the Treasury, would be making use of the survey in considering how risk management in departments might be further strengthened and internal risk management processes kept under review. ILGRA was planning to produce some guidelines for departments, by 2002, on risk management, including how to tackle risk issues that are likely to be amplified or attenuated by the media; and
- the project would be implementation-focused and would seek to improve the take up of risk management best practice in the public sector. It would help to inform the Government's Statement on Risk.

The preparation of this Statement on Risk, which had been proposed by the Cabinet Office Better Regulation Unit and ILGRA, was also described in the same Government Response, as follows:

> The Government will soon publish a Statement on Risk. The Statement will set out the principles of good risk management. It will also frame further work ... particularly focused on improving the coherence and consistency of risk management across the Government's activities.

The further work was to be undertaken by the Performance and Innovation Unit (now the Strategy Unit) of the Cabinet Office and would look

> at ways to achieve more effective management of – risks to the public, including risks to public health, safety and environmental risks; risks to the conduct of the Government's

own business; and areas where some or all of the responsibility for risk falls to the private sector or individuals. The primary focus is on how to ensure that risk management is properly embedded into departments' decision making processes.

The end result was the publication of the Statement on Risk in the form of the Strategy Unit report in 2002 bringing to completion the preparation of the common framework towards which ILGRA had been working.

Finally, on the question about the current situation, it was announced in the Strategy Unit report that the work of ILGRA would be absorbed into the role of a new Implementation Steering Group "to drive change over the two-year period leading into the next Spending Review (2004)". The Group would report to a Ministerial Committee and to the Civil Service Management Board. An interesting question in the light of the discussion in Chapter 1, but one that must be left for another day, is whether this reorientation to a strongly vertical form of integration from the largely horizontal integration pursued by ILGRA will be more effective. It will be recalled that a conclusion of one of its own commissioned studies was that ILGRA in its existing form had limited potential for achieving reform, lacking authority or sanctions. It is apposite to wonder how an endeavour "to drive change" – with a power to apply financial sanctions and hence the attendant risk of concentration of effort by the actors on their avoidance – in a complex system can be compatible with the systems thinking that has been reported (*Observer*, 2002) as coming into vogue in government.

Concluding Remarks

In this chapter, the experience of the ILGRA network has been used as a case study to draw lessons pertinent to the aims of this book as described in Chapter 1. The case study illustrates the operation of a largely horizontal mode of integration. The key actors were individual departments motivated by a need for consistency – though not uniformity – in their approach to policy-making on risk. These policies have to be discriminatory since the significance of a risk varies between policy domains. The case study also illustrates how other policy areas – science and technology, communication, microeconomics – were brought appropriately into the framework for integration through the need to improve the knowledge base and to develop principles and methodologies. Furthermore, the whole endeavour was anchored in high level determination of overall policy objectives.

By its nature, the work of ILGRA and its successors is never done. The issue of risk is one that will continue to present new challenges to government policy-making and particularly to the achievement of joined up working. Although it is not the place of this case study to elaborate further on reasons for this and the elements that it would comprise, nonetheless it may be remarked that the outstanding issue that will beset the development of policy is the emergence of what has been dubbed as "societal concerns". The term, first adopted by HSE (HSE, 2001), is used to describe society's, as distinct from an individual's, aversion to particular forms of risk, for example multi-fatality accidents or, more evocatively, risks from hazards that are perceived to pose a threat in some way or other to the very basis of society as currently structured and which readily form the basis of media scares such as "Frankenstein" foods or the "grey goo" of nanotechnologies. There is an evident need for consistency in the handling of such issues and the parameters that matter in their resolution – especially the role of trust in institutions – are yet to be agreed.

In relation to the governance debate as described in Chapter 1, the case study illustrates how the ILGRA network adopted a number of mechanisms to bring non-government actors – mainly from academia and consultancies though with some from industry and NGOs – into the deliberations on methodological issues and generally to increase the transparency of policy development. As a result, there was a considerable enrichment of the interactions between these actors and the government network (and also it may be said between these actors themselves). Thus, the role of ILGRA in governance terms was in facilitating this interaction. The network itself did not have non-government interests directly involved in its business meetings. It has been suggested that such involvement would have brought benefits – especially in relation to unravelling the issue of societal concerns just mentioned. However, that must remain a matter of debate and all that can be said is that ILGRA would have been a different animal and this case study would have taken a different course.

Acknowledgement

The work described in this chapter resulted from the contributions of the many people who took part in ILGRA and its subgroups. The author is grateful for their co-operation and encouragement during his period as chairman from 1992 to 1999. Responsibility for the views expressed and for any errors rests with the author alone.

References

Chapman, J. (2002), *System Failure: Why Governments Must Learn to Think Differently*, London: Demos.

Chilton, S. et al. (2000), *The Valuation of Benefits of Health and Safety Control*, Report CRR 273/2000, Sudbury, UK: HSE Books.

DEFRA (2001), *Response to the Report of the BSE Inquiry*, Cm 5263, London: The Stationery Office.

Gillies, J.M. and McCarthy, I.P. (2002), 'Adaptable Networks – Perspectives from a Business Context', in James, R. and Miles, A. (eds), *Managed Care Networks: Principles and Practice*, UK Key Advances in Clinical Practice Series, London: Barts and The London, pp.65-90.

HM Treasury (1996), *The Setting of Safety Standards*, Report by an Interdepartmental Group and External Advisers, London: HM Treasury.

HMSO (1990), *This Common Inheritance: Britain's Environmental Strategy*, Cm 1200, London: HM Stationery Office.

HMSO (1992), *This Common Inheritance: The Second Year Report*, Cm 2068, London: HM Stationery Office.

HMSO (1995), *Forward Look of Government Funded Science, Engineering and Technology*, London: HM Stationery Office.

Hood, C., Rothstein, H. and Baldwin, R. (2001), *The Government of Risk: Understanding Risk Regulatory Regimes*, Oxford: Oxford University Press.

House of Lords (2000), *Science and Society*, HL Paper 38, London: The Stationery Office.

HSE (1998), *Developing a Common UK Approach to Negotiations on Risk Assessment at International Level*, Report by National Economic Research Associates, London: Risk Assessment Policy Unit, HSE.

HSE (2001), *Reducing Risks, Protecting People*, Sudbury, UK: HSE Books.

IEH (1999), *From Risk Assessment to Risk Management: Dealing with Uncertainty*, Leicester, UK: Institute for Environment and Health.

ILGRA (1996), *Use of Risk Assessment within Government Departments*, Sudbury, UK: HSE Books.

ILGRA (1998), *Risk Assessment and Risk Management*, 2nd Report of ILGRA, Sudbury, UK: HSE Books.

McQuaid, J. and Le Guen, J.M. (1998), 'The Use of Risk Assessment in Government Departments', in Hester, R.E. and Harrison, R.M. (eds), *Risk Assessment and Risk Management, Issues in Environmental Science and Technology*, No. 9, Cambridge: Royal Society of Chemistry.

Observer (2002), 'Thinking outside the Box', London: *The Observer*, 26 May 2002.

Phillips (Lord), Bridgeman J. and Ferguson-Smith, M. (2000), *The BSE Inquiry, Report No. 887-I*, London: The Stationery Office.

Strategy Unit (2002), *Risk: Improving Government's Capability to Handle Risk and Uncertainty*, London: Cabinet Office.

Waldegrave, W. (1987), *Sustaining the Environment in a Developing World*, NERC Annual Lecture, Swindon, UK: Natural Environment Research Council.

Chapter 6

Rural Policy: A Highlands and Islands Perspective

Frank Rennie

There is a special challenge in looking at changing issues of policy and governance from a rural perspective. In particular, the aspirations for a sustainable development policy that is comprehensive, integrated, and inclusive, seems to cover every aspect of life, science, and society, and to change with each individual usage, but first I need to sketch what I mean by "sustainable". There are plenty of definitions and measurable indicators currently available (DOE, 1996; Clayton and Radcliffe, 1997; UNDESA, 1998; Hart, 2000) but here I am not concerned with splitting hairs over definitions, or even describing the scope and scale of the problem. There are any number of articles on these issues. What I am interested in at present is the need for rural development theories to become more sophisticated, more holistic, and more balanced in their mix of quantitative and qualitative measurement of sustainable rural development for the forthcoming decades. In this respect I want to simplify "sustainable development" to four major components, and to look at the main elements that constitute these components in relation to policy and governance in the context of the Scottish Highlands and Islands.

The four main components of sustainable development, I would postulate, are:

1. Environmental sustainability;
2. Social sustainability;
3. Economic sustainability;
4. Social equity.

Before we dissect each of these components in turn, we should spend a little time considering the policy drivers that run through each of these four components. Policy drivers are the push-pull horizontal forces that dictate events at the multi-sectoral level, in many cases at the multi-national scale, and because of this they are frequently conflicting, cross-disciplinary, and flatly contradictory. On a sliding scale of global to local, the main policy drivers can be considered to be:

- Globalisation and trade liberalisation;
- European Union Directives;
- Westminster Parliament decisions;

- Scottish Parliament decisions;
- Local authority, business, and community demands.

The interaction between these policy drivers and the practical results in rural areas can be considered to be dynamic, only partially predictable, and part of a complex adaptive system that is constantly changing and evolving (Marten, 2001). In keeping with similar complex adaptive systems, the emergent properties at each level, or stage, of decision-making are likely to be present in different combinations and with an emphasis on different priorities, and each level may be different (but complementary) to the whole system. Understanding this flexibility and complexity of the policy ecosystem (and the analogy with the biological equivalent is deliberate) is fundamental to our appreciation and understanding on how to react, plan, and evaluate how the changing policy landscape actually impacts on rural reality.

It may seem strange to single out sustainable development issues at this stage in the discussion, but bear with me. Although there has been general agreement about the four main areas of "sustainable development", the indicators by which the progress towards sustainability can be measured is less certain. It is important to realise that we are seeking to move towards sustainability in development, rather than to reach some ultimate, fixed target where "sustainability" has been achieved and we can all sit back and relax. The problem, to be specific, is not a lack of indicators that we can use to measure the various elements of sustainability, but a lack of agreement on which indicators are the most appropriate for any given sector, scale, and geographical region.

We are almost literally awash with indicators, but most of them lack any universal applicability. In order to measure progress it depends upon where you start measuring and in what general direction you want to travel. In order to develop policies to improve the effectiveness of governance, it is necessary to have some general (even minority) consensus of what we want to improve towards. This is far from clear at the present time.

We have only recently emerged from an era when sustainable development was regarded as almost synonymous with sustainability in the maintenance of environmental conditions and habitats. As society has come to appreciate that many of our ideas of "good" environment are socially constructed, we have developed a new understanding to include the need for our policies and actions to be more socially and economically sustainable. More recently, we have included the need to strive for greater social equity as we have realised that, for example, the presence of sustainable services such as education, health centres, or shopping outlets may be available, but not to everyone in society.

Environmental Sustainability

A common concern, among both agency professionals as well as the "ordinary" public, is the overlapping bureaucracy relating to environmental protection. In particular, the perception is of layers of designation, one piled on top of another with promiscuous overlap, specialisation, and in/exclusion of targeted protection, lost in a sea of acronyms and abbreviations. We have SSSI (Sites of Special Scientific Importance) that are also NNR (National Nature Reserves) and SSA (Special Scenic Areas) or SAC (Special Areas of Conservation) some of which might also have RAMSAR (wetland) protection, ESA (Environmentally Sensitive Area) assistance, or have the epithet of World Heritage Site or World Biosphere Reserve. On top of that, where does Natura 2000, National Parks, or the local planning regulations fit into this jigsaw? It is difficult not to have sympathy for the farmer or countryside walker who gets lost in the acronym maze and simply retreats in frustration behind a barrier of apathy ("What does it have to do with me?") or hostility ("It is a distraction to me and I don't like it!") Few countryside users, whether they are based in the country or in the city, would reject the idea that society needs to have some legislation to safeguard the natural environment, but it seems that even fewer can clearly explain the plethora of designations and what these controls actually mean, or do not mean, on the ground.

Against this patchwork (or should it be papier-mâché?) framework, the policy formulation needs to be contextualised against the current period of rapid agri-environmental changes. The period since the end of the Second World War has seen massive changes in the structure of agriculture and the way agricultural activities are supported. Much has already been written about the loss of agricultural labour, the trend towards bigger, more specialist farms, and the costs, benefits, and concerns over the Common Agricultural Policy. These are very real concerns, not just for those people who have been making their living from agricultural production (or its supply and processing aspects) but for all of society.

The growing public attention that is paid to the origin of food products is bringing these policy issues into the home of every consumer, and that means every person. At varying times and with varying degrees of public profile, we are made aware of the debates on organic foods, genetically modified crops, animal disease in the food chain (BSE, foot and mouth, chicken flu) and the politics of ethically fair trading with less developed nations. The debate has moved on from the post-war farm shortages when the drive was simply to increase food security by increasing food production, and we are now beginning to specify on a global scale what sort of food we want, how it should be produced/processed, and how this affects the lives of the producers.

This complex and evolving backdrop is difficult to pin down, even at the European or national level. Enlargement of the EU to include Central and Eastern European countries will undoubtedly skew attention and resources away from the rural areas further to the west of Europe that have been in receipt of significant financial support over the past decade or two. The exact shape of a set of reformed EU policies for European rural areas is still to be decided, the application is increasingly problematic, and the results are even less predictable than previously. It is little wonder that there is a switch away from direct support to agriculture, in favour of wider benefits for rural society, and increasing speculation on the benefits of following the route taken by New Zealand to phase out direct agricultural support entirely. There is a stark contradiction in Scotland/UK between the realisation of (and weaning away from) dependence of the countryside on CAP support (and arguably wider EU funding also) and the fixation of dealing with the countryside in compartmental terms (farming, forestry, housing, environmentalism). This contradiction filters back to the research and policy level, with, on the one hand, businesses (including farmers) being urged to be more entrepreneurial and innovative, yet, on the other hand, failing to maximise EU funding as additional funding that would enable streamlining and greater co-ordination at regional and national levels.

At a national and regional level, land reform has appeared back on the policy agenda after over 100 years of grumbling neglect. Although it seems to us to be a particularly Scottish affliction, historical circumstances of the Highlands and Islands (especially notions of the Highland Clearances) have combined with unfinished political business, and a new understanding of community empowerment with dramatic effects. There are two related issues, but they should not be conflated. First, the development of policies that would enable the ownership and management of land (and other natural resources) by the local community and, secondly, enabling reasonable access to the countryside for all responsible citizens.

The first issue really stole centre-stage with the purchase of the Assynt Estate[1] by the local community. This was a very public and emotive campaign, and though there was occasional hyperbole, the event was a momentous accomplishment for rural and community development. Circulating around a specific issue, the purchase of the estate by its inhabitants, there was a great movement of public and private goodwill that encouraged agencies such as Highlands and Islands Enterprise and Scottish Natural Heritage to invest significantly in a radical vision. Current policies were interpreted, not in terms of the nationalisation (public-ownership) of assets, or in "subsidy" to prop up a failing economy, but as an investment in a new type of organisation to promote sustainable community development. Enabling this local development trust to own and manage the land-assets that they occupied was attractive in economic, social, environmental, and political terms. It was seen as a fundamental signal from government to establish a template that would enable other communities to make and implement their own development decisions at a local level.

Since that event there has been a steady trickle of communities, large and small, that have successfully bought out their own community assets, and as the concept becomes more entrenched in reality, eyes are turning internationally to study other forms of community ownership of natural resources, such as forests, freshwater resources, and other types of natural capital. Also in response to this new-found optimism in community ownership and management, the Scottish Parliament has debated and passed legislation to extend the rights of community ownership, as well as to make the process of acquisition easier, a clear case of

---

[1] See, for example, http://www.scotland.gov.uk/News/Releases/2003/08/3953 (accessed 8 September 2004).

opportune populism stimulating national policy formulation from the grassroots upwards, in this case quite literally.

In contrast, the debate about access to the countryside has thrown into sharp focus the contradictions and divergence of views over what people think the countryside should be used for. The growing strength of the "environmental movement", together with even a mild "greening" of politics has prompted society to consider the obligations of wider issues of cause and effect. This extends from a greater awareness of the symbiotic relationship between town and country (often, however, to the detriment of the rural areas) to the growing unease with the creation of "islands" of nature reserve amid "seas" of under-protected land.

The former manifests itself in the developed countries as increasing numbers of city and town dwellers turning to the countryside for their recreation, with increasing demands upon rural tourism, roads, cultures, and housing. Policy is a blunt instrument to deal with this directly, although higher taxation on second homes and the creation of "honey-pots" such as National Parks to attract and contain mass vacationers are the most obvious direct side-effects. The latter issue may be easier to deal with, bringing us back to the layers of designated protection afforded to some areas of high-value nature conservation while, adjacent to these, habitat of apparently similar quality is not afforded such protection.

The nature conservation obligation to consider "the wider countryside" is becoming more pervasive, in issues such as planning blight, landfill sites, and the location of turbines for wind-power generation. Furthermore, there is no likelihood of the situation improving without a radical overhaul of public expectations of what we actually want to use the countryside for – the combination of recreation, food production, and other natural resources. This will not happen without a clearer appreciation of the role of social sustainability and how this combines with the environmental management issues outlined here to shape our expectations of policies for appropriate governance.

Social Sustainability

Social sustainability has enlarged its scope, from being primarily concerned to redress rural depopulation (though this is still a major concern in many European rural areas) to include a much broader range of societal structures and services. Social sustainability in rural areas has come to include almost every aspect of society, from the provision of a viable local primary school to the opportunities for access to higher education by distance- and online-learning, from the quality of health services locally available, to the vitality and diversity of the arts, music, and social life.

This enormous potential scope, and the flexibility of the precise definition, means that there has tended to be a confusion of overlap between different agendas. Throughout the region, the Highlands and Islands Development Board (HIDB), subsequently replaced by the enterprise company network (HIE), has long cherished the fact that its remit includes a statutory requirement to consider the social development of the Highlands and Islands communities. This remit has not been gifted to the Scottish Enterprise Network in the rest of the country, despite some cogent arguments by spokespeople from the Borders and other rural areas that it should be. In numerical terms, financial spending on social development has always been a tiny percentage of total HIDB/HIE spending on regional development, certainly always in single figures, but there are innumerable anecdotes and case studies to provide evidence that the community value of the investments is rated considerably higher than the mere value of the financial contribution from the agency.

Tensions have arisen from time to time between the local authorities and the local enterprise companies regarding who should be primarily responsible for social (and economic) development, and while these tensions have largely been resolved amicably on a case-by-case basis, there is a groundswell of agreement that there is an inbuilt potential for confusion and duplication of effort. In an attempt to respond to international initiatives to increase the level of community participation in the formulation and management of sustainable development (for example, Local Agenda 21, see Lafferty, 2001), numerous policy directives have spawned from national government. A number of these have the stated aim of fostering multi-agency and/or multi-sectoral co-operation at the regional or local level, such as the consultation committee for the Community Planning Framework, or for the Local Economic Forum, or, to a lesser extent, the Local Biodiversity Action Plan (LBAP). The verdict on the effectiveness of these initiatives has been mixed, with some rural areas, particularly smaller areas of governance, reporting that similar issues of policy and planning are being discussed in different committees by a small number of "the usual suspects" wearing different combinations of hats.

Finally, it remains to be seen what the long-term effect will be of the perceived recent trend for central government to shift from a close involvement in the delivery of direct services, to a more strategic role of policy formulation and service management facilitation. The provision of funding under a service level agreement (effectively a franchise from government) instead of providing a state-run service has actively encouraged many local and community-active organisations to provide a whole new range of local/regional services. The issues are complex, being driven on one hand by a wish to shrug off the hand of a "nanny state" and encourage local empowerment but on the other to ensure that the state does not abrogate responsibility for social service provision in rural areas as a convenient side-benefit to cost cutting.

In general, the greater participatory involvement of local communities in the provision and management of local social services has been positive, and has led to both innovation and local customisation of service provision. Concerns remain, however, that the fragmentation of service delivery will produce variable quality in different geographic areas, or that gaps will emerge, particularly in rural areas where sparse population and higher structural costs will mitigate against private sector profits (or produce market failure unless different funding levels and/or standards are applied).

Economic Sustainability

Implicit in the current rhetoric for sustainable development is the imperative for financial sustainability. This is nothing new, of course, except that with an entire generation having grown up under the subsidised agricultural production of the European Union Common Agricultural Policy (CAP), and the growing reaction against this in recent years, it is important to search for a balanced perspective. Critical to this perspective is an understanding that many rural areas are not capable of achieving a completely internal financial self-sufficiency, and may never be so, due to the characterising factors of small, thinly distributed populations, geographical difficulties with service provision, and organisational dis-economies of scale.

In the global drift towards "free" markets there are large forces at work that are altering and distorting national and regional trading patterns, consequently economic sustainability, at both the local and the global levels, is volatile and the details are unpredictable. This tends to create confusion, frustration, and mistrust in the vertical process of policy implementation between the "big picture" macro view of government, and the immediate realities of the micro-economic decisions made at community level.

From the rural perspective there are tensional forces that simultaneously encourage (1) local communities and businesses to become more engaged with the global economy, mainly, but not entirely, through a fuller participation in the exchange of digital data over the internet; and (2) to prevent financial leakage in the local economy by greater efforts to seek self-sufficiency in commodities that can be provided locally, such as agricultural products, public services, and labour. These tensions are by no means mutually exclusive, but there is a sliding scale of intermingling, and the complexities of dynamic change mean that it may take a long time before any equilibrium position is approachable.

Rural businesses and agencies that seek to accommodate these changes in a strategic manner, by personalising their global/local profile in a manner that I have called "localisation" will be best placed to survive in the 21st century. Combinations such as hotels/guest-houses that simultaneously advertise and do their business transactions online, while utilising local specialities of food, culture, and local environment, will be in a position to reap the best of both policy worlds.

Other areas, such as e-learning, provide both opportunities and threats for rural areas – opportunities for remote learners to access educational providers more easily, or for rural colleges/research centres to offer their portfolio of activities on a global stage, but threats because local learners may decide not to attend the local college, or because the global reach of education means that other colleges/universities can now significantly extend their traditional catchment areas (Bryden et al., 1996). For rural areas in particular, the concept of digital inclusion is not merely about whether or not an individual is enabled to access the hardware and software to take part in internet activities, but also in what manner they take part, i.e. as active participants and decision-makers or as simple consumers of urban commodities and policies (Ascherson, 2003).

As in the arena of social policy, the trend in policies for economic development in rural areas is changing from simple financial support to more advice, support, and training, and financial assistance is more clearly targeted than in previous decades. Part of the response to try to make State financial support go further has been the application of the now almost ubiquitous condition of match funding. In many sectors this has moved beyond the reliance of targeted financial assistance from the European Union (through ESF, ERDG, LEADER, etc.) to include a range of innovative partnerships with both public and private sector organisations. This includes not only "sweat equity" and contributions in kind (labour and/or consumables) but also joint or franchised property developments such as advance units on industrial estates, or more controversially constructions like the Skye Bridge or local school premises, constructed by companies with primarily a profit-making motive. Whether these continue to be viable funding options for public construction in rural areas remains to be seen, though it is difficult to see that there could be a great deal of private interest in constructing buildings in the not-for-profit sector, especially in geographical regions where the costs of construction can be 145 per cent of the average cost of similar premises in the Central Belt.

Finally, the future of farming remains uncertain. Against a constant backdrop of change since the First World War, we have seen the steady decline of farm labour and farm incomes while farm sizes and outputs both continue to rise. The policy and practice has now gone beyond the stage that is of interest/concern merely to farmers, and a steady stream of special-interests is forming up to demand a greater say in rural policy formulation and implementation. Environmentalists, countryside ramblers, a whole range of other outdoor recreational activities, heritage societies, town-dwellers with a view to what they would like to see in the countryside, urban shoppers who want to be able to trace where their supermarket purchases originated, specialist growers, organic food enthusiasts, or protesters against genetically modified food, all have a legitimate interest in what our society does with the countryside.

This spreading constituency presents opportunities for rural policy-makers to bring rural issues more centre-stage in governance, more mainstream in cause-and-effect, but also creates the problem of trying to establish a common vision for the countryside. With so many different, overlapping, and contradictory interest groups, all set against a constantly shifting policy regime, it is becoming harder to establish a consensus of what policy is actually *for* – what are we trying to achieve? How will we know when we get there?

Policy shifts of recent years, such as access to the countryside legislation (freedom to roam) and land reform (the community right-to-own) have been seen as radical, and indeed they are, against the backdrop of centuries-old marginalisation of rural political issues, but in order to harness a common vision and put them at the centre of democratic government, we need to produce something much more radical. Such a rural policy not only steps from the margins to the mainstream of government business, but also provides the fulcrum for a much wider range of policies that affect our whole society, decentralised quality housing, the deployment of a digital communications infrastructure for the 21st century, re-location of employment from over-crowded, economically over-heated cities, and much, much more.

The rural debate has moved far beyond the immediate concerns of how much food farmers can produce, to encompass fundamental questions about how we envision our society to be, urban *and* rural, how we ensure a sustainable quality of life for all citizens, and what role does the rural environment play in this bigger picture?

Social Equity

This is an enormously important area of policy formulation frequently overlooked by proponents of sustainable development, but as important, I would suggest, as the three traditional concerns of environmental, social, and economic development. The reasoning is like this: there is little point in pretending to achieve sustainability in other areas if these achievements are unequally distributed to an elite few of the population. Having the highest GDP, the best standards of university education, or the presence of world-class hospitals is a hollow achievement if the benefits are denied to the majority, or even a significant minority, of the population. For this reason we correctly have laws that demand that public policies should be subjected to social inclusion proofing, gender proofing, and check-listed against other forms of discrimination, accidental and intentional. Be that as it may, there is still a strong tendency to impose urban solutions on rural problems, or at least to overlook the need for *different* solutions to rural problems in order to maintain a measure of social equity across the country.

Whether we are considering the Health Service, local housing reforms and quality improvements, or the delivery of education, there is a recurrent danger that reforms to the systems neglect the rural context. In the drive for equality, the focus is too frequently on the equalisation of the resource base, not on the equalisation of the outcomes. Again we return to the issue of creating a vision with national consensus. How do we ensure that all citizens have an equal entitlement to a high quality of life that is not based on the random geographical distribution by birth?

Current attempts to compensate for socio-economic exclusion based upon indicators tied to identification by national postcode are but a crude measurement of dis/advantage. In Scotland we lack a national co-ordinating structure that attempts to integrate issues of rural interest vertically (community   government) and horizontally (between sectors). We have sectoral efforts, and a few localised cross-sectoral fora, but the notion of "rural proofing" governance, in the way that it is scrutinised for compliance with gender and equal opportunities issues, is noticeably absent.

Trends

Despite the complex policy landscape, there are a number of clear, generalised trends that can help to inform our vision of what rural policy-makers should be considering as priority issues. First, the "traditional" divide between indigenous country dwellers and generation after generation of families living in the town is being broken down. As the mix of backgrounds and lifestyles becomes more heterogeneous, so do the expectations and demands of rural living. I don't mean by this that standards in education, health, etc. are becoming more stringent (though this may be so) but simply that there is a greater diversity in what families consider "appropriate" and a greater expectation that they should not be forced to suffer public services worse than their urban counterparts simply because of their geographical location in the country.

This has led to increasing demands for distributed access to services, peripatetic professionals, innovative solutions to provide specialist cover despite low population numbers, and 24/7 access with helplines, e-learning, and other online services being part of the response. There is concern that "remote access" might encourage further remoteness – medical consultation provided at a distance, local government information over a phone or Web site, both with minimal face-to-face contact with individuals – and though we will undoubtedly see much more of this in the future, it is clear that a balance needs to be struck between pervasive access and social presence.

Following on from this, we have seen a trend for administrative services to become centralised, but distributed, i.e. the function is centralised but the location need not (as in the past) be based in a major town or city. Such re-location is not always based upon high technology solutions, but improved access to networked computer power frequently underpins these activities, and ubiquitous access to broadband internet will only provide a stronger driver to support re-location. There are many other reasons to support re-location of business interests to rural areas of course, and quality of family life, reduced commuting time, better transport links, reduced costs of premises and operational overheads all come into the calculation, but the ability to access affordable, high-quality internet connections easily will be a major deciding factor. As a result we will see "broadband" policies for global networking becoming a major policy issue at all levels, from national government, through local authorities, to the proverbial "man at the farm gate".

Linked with this, although it is in fact a much broader subject area, is the need to recognise (and capitalise) on regional trends for greater localisation and local branding, and niche marketing. In a general sense, the internet has allowed and encouraged even small local businesses to do their advertising and business transactions on a global scale, this is only a small part of the picture. Within the European Union, for instance, we see inexorable political moves towards greater international unity, for example, commonality of laws, sharing of intelligence and defence, a common currency, and the strengthening of English as the omnipresent business language. We also observe a resurgence of the importance of regional culture and identities. While political homogenisation may make for greater efficiency and blandness, the assertion of divergent cultures and customs provides for a celebration of the uniqueness and individuality of regional identities within a common European framework.

In this age of almost uniform urban cultural mix, the most distinctive regional cultural identities are those that are derived from the rural lifestyles of the regions. This will have a potentially huge impact for gaining local attention in a high-pressure global tourism market-place, and affects not simply hotels and travel firms, but the production of regional music, arts, literature, food delicacies, wildlife, lifestyles, and ultimately how regions are able to reinforce themselves with a positive self-image. This is an area of community life that spans the full gambit from natural science to the arts, it is technology-based but not technology driven, and is a multi-million pound investment that is poorly tracked due to overlapping and crosscutting policy agendas. Perhaps it is time to reverse the conventional flow of policy-making decisions and start with the grassroot expectations of demands and opportunities.

Part of the solution may be linked to the already mentioned drive to reduce the involvement of the State in public service provision or new infrastructure. An extension of the greater private interest group involvement (for example, public-private sector partnerships for the construction of new public buildings) might include new community-level partnerships for the construction of locally/regionally appropriate public policies. This need not be counter to the imperative for nationally appropriate policies, but would provide a level of customisation and personalisation in the interpretation and implementation according to the consensus of identified local priorities.

Explicit in almost every sector of governance, and indeed of our wider societal structures, are the escalating demands for greater online service provision. These can be read everywhere, for example:

- the prevailing inclusion of e-learning in the blends of learning styles offered to learners (initially primarily remote and rural, but now increasingly cosmopolitan);
- the desire to make available on the web information relating to our local authorities, health boards, development companies, and other public bodies (previously held as paper copies for limited circulation) partly as cost-saving, partly for greater transparency;
- the drive for even very small "micro-businesses" to get online in order to advertise the products, gain new customers, collaborate with others to win economies of scale, and so on.

The web is a new medium, complex, constantly adapting, and fast changing, so it is difficult to predict any change with certainty, other than that access will escalate rapidly, new uses will be developed to meet specific needs, and that this growth will continue to be unpredictable. The implications of rural non-involvement in the electronic revolution are stark and uncompromising, and unlike the industrial revolution they will have immediate and negative effects on all citizens. Despite this we are barely beginning to take seriously the policy obligations for governance.

Dangers and Opportunities

It would be foolish and disingenuous to claim that these changes will all be for the best. There is a down side to every new technology, and every policy directive has its gaps, exclusions, and time-limitations. The need to move from representative governance to more participatory structures, from policy issues that are solely top-down to those that reflect local and regional priorities within a national and international framework, will not be easy. I am tempted to say that for a while, in the settling-out process, we will be "buried under paperwork" but even this well-clichéd phrase is redundant – the bureaucracy will continue but in the new electronic age much of it may never appear anywhere on paper. With the progressive and speedy shift to the currency of digital information, there will come a certain loss of local autonomy, and this co-responsibility is an inevitable consequence of creating wider partnerships. I have tried to suggest that this could be more than compensated for by an increase in localisation of the public sector and business agencies that will deliver a more appropriate, integrated, locally validated set of policies that are supportive of sustainable regional development.

This is the goal, but the path will prove to be laborious in the short-term, with the threat of death (or petrification) by serial consultations and bureaucratic consideration by committee. A key feature in any complex adaptive system (for this is what the policy environment is) is that every bit of the network is potentially linked to every other bit, so when we alter one bit, the ramifications can be extensive and unpredictable. In the context of rural policies, we will need initially to bring all the potential participants to sit round the same table, map out the links and areas of mutual interest, and then allocate operational areas for action within agreed frameworks of self-reference and collaboration.

Although the potential linkages are complex, new technology gives us new tools for ensuring transparency and feedback mechanisms that enable wide participation and popular involvement in the process of fine-tuning the policies and practice of governance to appropriate levels. The alternative is to run the danger of (or, some would argue, persist with) progressive homogenisation of bureaucratic procedures while persisting with fragmentation of interest groups, participants, and respondents. Policy will become (even more) amorphous rather than integrated; provider-led rather than user-centric; and inward-looking rather than aspire to wholesome visions of collective public good. The fact is that new communications and working practices are shrinking distance and therefore in many aspects are blurring the distinction between urban and rural societies. Despite this, there appears to be a disparity in the weight of opinion given to the credibility of scientific data versus social considerations. The force of science and technology as a major policy driver in agriculture is no longer a main priority (though GM crops might raise this) and more sophisticated debate on the role of human interaction with the landscape is now common. In this respect it is interesting to note that the impetus for renewable energy generation in rural areas (especially wind turbines) may actually succeed in uniting the hitherto opposing factions of the "environmental" lobby and the "community organisations" social agenda in common opposition to perceived inappropriate development activities.

Opportunities exist for rationalisation of public services, for example, but these may be distributed strategically in the interests of user efficiency rather than simply in response to financial cost cutting. One-stop-shops, distributed public sector work-stations, remote working, and shared resources may offer collaborative solutions for enhancing the value of public goods and social capital, but it will require greater public awareness and participation in the decision-making machinery. With the perceived lack of interest in the party political agenda, the impetus for this will only come from a radical overhaul of the ways in which we seek to secure the engagement of individuals and local communities in a manner that clearly demonstrates added value to the process of becoming involved. Area marketing and regional niche products offer exciting opportunities to link vertical distribution with horizontal production and area branding, for example, to establish the Outer Hebrides as a region of distinctive heritage values, linking the clear coastal waters (fishing, surfing, diving) with high quality seafood, and linking this with other locally-produced food, good local restaurants where visitors can hear high-quality traditional music and Gaelic song. These linked "packages" will not be solely for the tourist, but also for local use and interaction. The ease of internet links will not limit these initiatives solely to the local market, but will also encourage global export opportunities, with each geographical area having an identifiable brand of local distinction to promote on the world stage.

There are two key cornerstones of this new policy landscape, and they are inseparable. One is the so-called "digital agenda" – the new age of knowledge economy that pervades education, business, and social dealings – and the other is the democratic agenda, that includes fostering greater public participation in governance, greater transparency, and innovative ways of addressing common responsibilities, such as community stewardship of land and natural resources. In the complex nature of shifting social and scientific policies it is hard to say which agenda should be addressed first, but it is safe to say that the future promises to be digital "chips with everything".

Conclusions

In terms of my initial analysis, there are five clear conclusions for the changing issues of policy and governance from a rural perspective:

1. The need for simplification, integration, and re-prioritisation;
2. The need for clear public policy guidelines that demonstrate the costs and the benefits of involvement and the imperative of local participation;
3. The need for transparency in government, not only in decision-making, but in the inter-linkages between sectors and policies;

4. The need for rapid access to (digital) information for education, collaboration, and the dissemination of information will dramatically alter the future policy landscape;
5. The amount of re-thinking the modes of policy integration in order to ensure both vertical integration (between central government and the peripheral regions) and horizontal integration between all policy areas that impact upon rural areas is a huge, and as yet barely addressed political challenge, but one that is central to the analysis of governance in the next few decades.

References

Ascherson, N. (2003), 'Designing Virtual Citizens: Some Scottish experiments with electronic democracy', *Scottish Affairs*, 43 (Spring).

Bryden, J., Fuller, T. and Rennie, F. (1996), *Implications of the Information Highway for Rural Development and Education*, Report of the Arkleton Trust Seminar, Douneside, Aberdeenshire, Scotland, February 1995, Enstone: The Arkleton Trust.

Clayton, A.M.H. and Radcliffe, N.J. (1997), *Sustainability: A systems approach*, London: Earthscan.

DOE (Department of the Environment) (1996), *Indicators of Sustainable Development for the United Kingdom*, London: HMSO.

Hart, M. (2000), *Sustainable Measures*. URL: http://www.sustainablemeasures.com/ (Accessed 11 February, 2004.)

Lafferty, W.M. (ed.) (2001), *Sustainable Communities in Europe*, London: Earthscan.

Marten, G.G. (2001), *Human Ecology: Basic concepts for sustainable development*, London: Earthscan.

UNDESA (United Nations Department of Economic and Social Affairs) (1998), *Measuring changes in consumption and production patterns*, 83pp URL: gopher://gopher.un.org/ 00/esc/cn17/1997-98/pattern/mccpp5-9.txt (Accessed 11 February, 2004.)

PART III
THE LIMITS TO INTEGRATION

Chapter 7

Hypermobility: A Challenge to Governance

John Adams

The year 2001 set a record for new motor vehicle sales in Britain – 3,137,700 were sold. The following year surpassed this record with 3,229,400. 2003 produced yet another record – 3,231,900. The forecast for 2004, at the time of writing, is a number of a similar size – another record or nearly so.

Allowing for the scrapping of old vehicles, the annual increase over these years in Britain's motor vehicle population has been over 800,000. Allowing 20 feet for each of these vehicles (the distance between parking meters) we can estimate the size of the parking space they demand – each year the equivalent of a new car park stretching from London to Edinburgh more than nine lanes wide was required to provide one parking space for each of these extra vehicles.

The threat posed to the natural environment by the growth globally in the numbers and use of cars, and the much more rapid growth of air travel, has received much attention but little effective action. The social consequences of this growth have received much less attention. This growth, I argue below, is not only exacerbating environmental and social problems, but making solutions to these problems by democratic means – by governance[1] – increasingly difficult.

Policy-makers sometimes claim that transport policy is more "joined up" than many other policy areas. However this claim usually refers to horizontal integration between infrastructure planning and transport policies. There are no effective mechanisms, horizontal or vertical, for integrating the wider societal impacts of hypermobility described here into planning and transport policy-making.

---

[1] The first meaning for "government" and "governance" in the dictionaries I have consulted indicate that the words are synonyms for "the act or process of governing; *specifically*: authoritative direction or control". But a distinction appears to be establishing itself between processes of governing that are top-down and those that are bottom-up. For the purposes of this essay I take *government* to refer to a centrally controlled top-down process, and *governance* to refer to a more bottom-up process for ordering human affairs characterised by more self-regulation and democratic accountability.

Hypermobility: Too Much of a Good Thing

Mobility is liberating and empowering but it is possible to have too much of a good thing. The growth in the numbers exercising their freedom and power is fouling the planet and jamming its arteries. Prodigious technological efforts are now being made to solve the problems of pollution and congestion caused by the growth of motorized mobility. Let us suppose that they succeed.

Suppose technologists were to succeed in inventing a pollution-free perpetual motion engine; the laws of physics dictate, of course, that they can never succeed, but this defines the goal towards which the motor industry and environmental regulators are striving. Suppose further that they succeed in developing the ultimate Intelligent Transport System – a computerised traffic control system that will hugely increase the capacity of existing roads, rails and airports. And finally, imagine a world in which computers are universally affordable and access to the internet is too cheap to meter; pollution-free electronic mobility is vigorously promoted as an important part of the solution to the problems caused by too much physical mobility. The lion's share of time, money and regulatory energies now being devoted to the pursuit of solutions to the problems caused by motorised travel is currently being spent on these "technical fixes".

To the extent that they succeed there will be further large increases in physical mobility. Cleaner and more efficient engines will weaken existing constraints on the growth of travel – either by making it cheaper, or by removing environmental reasons for restricting it. Intelligent Highway Systems promise to greatly reduce the time cost of travel by eliminating much of the time now lost to congestion. And electronic mobility, while capable of substituting for many physical journeys, is more likely to serve as a net stimulus to travel; by freeing tele-workers from the daily commute, it liberates them to join the exodus to the suburbs, and beyond, where journeys to shop, to school, to doctor, to library, to post office and to friends are all longer, and are mostly infeasible by public transport; and by fostering more social and business relationships in cyberspace it feeds the desire for "real" face-to-face encounters.

In 1950, the average Briton travelled about five miles a day. Now it is about 30 miles a day, and forecast to double by 2025 (see Figure 7.1). The growth trends for electronic mobility correlate strongly and positively with the trends for physical mobility, but their growth rates are much higher. Transport and communications provide the means by which everyone connects with everyone else in the world. The transformation – historical and projected – in the speed and reach of these means is having profound social consequences.

There are limits to what technology can do. A constraint on our behaviour that technology cannot relax is the number of hours in a day. As we spread ourselves ever wider, we must spread ourselves thinner. If we spend more time interacting with people at a distance, we must spend less time with those closer to home, and if we have contact with more people, we must devote less time and attention to each one. In small-scale pedestrian societies, *hypo*mobile societies, everyone knows everyone.

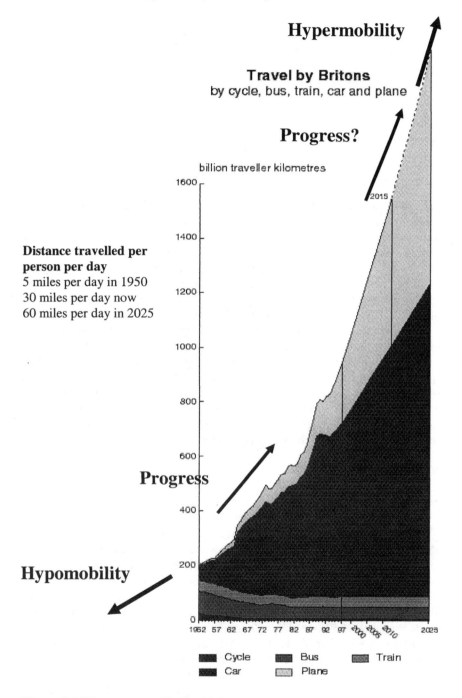

Figure 7.1 Distance travelled by Britons

In *hyper*mobile societies, old-fashioned, geographical communities are replaced by aspatial communities of interest – we spend more of our time, physically, in the midst of strangers. The advantages of mobility are heavily advertised; the disadvantages of *hyper*mobility receive much less attention. Many of the unwelcome characteristics of the hypermobile society can readily be imagined by extrapolating existing trends.

The Hypermobile Society

It Will Be More Dispersed

The process of suburban sprawl will continue. Societies whose members move at high speed over great distances consume more space. It is the long distance journeys – by road and air – that are experiencing the fastest growth rates. Walking and cycling – the local, healthy, democratic, and environmentally benign modes of travel – are in steep decline. Even with pollution-free, perpetual-motion engines there will be unwelcome environmental consequences. More of the country will need to be paved to provide parking places; the extra roads required will scar cherished landscapes and subdivide still further the habitats of endangered species; room will have to be found for new and larger airports; those parts of the world valued for their remote tranquillity will be further encroached upon. The London-to-Edinburgh car park referred to at the beginning of this chapter provides only one space for each extra vehicle. Their owners will want parking places at the other ends of their journeys, plus wider roads to get them there. High-tech "solutions" such as congestion charging aided by satellites and tracking devices will, in the absence of constraints on traffic growth, encourage further dispersal into areas where there is still room to move and park.

It Will Be More Polarised

The increase in the mobility of the *average* Briton described above conceals a growing gap between the mobility-rich and the mobility have-nots. All those too young, or old, or otherwise disqualified from driving will get left behind, along with those too poor to afford cars and plane tickets. They will become second-class citizens dependent for their mobility on the withered remains of public transport or the good-will of car owners. And as the world runs away from them to the suburbs most journeys will become too long to make by foot or cycle. Worldwide the mobility have-nots are still increasing. Despite a ten-fold increase in the world's car population since 1950 – to about 500 million – because of population increase, over this period the number of people who do not own cars has more than doubled – to about 6 billion. And despite the much more rapid increase in air travel over this period the number of people in the world who have never flown has also increased. In Britain, and worldwide, the onrushing trends are fostering a mobility apartheid.

It Will Be More Dangerous

For those not in cars there will be more metal (or carbon fibre) in motion. The increase in danger is not well reflected in accident statistics. The fact that there are now about one third as many children killed every year in road accidents as in 1922 when there was hardly any traffic and a nationwide 20mph speed limit, does not mean that the roads are now three times safer for children to play in; they have become so dangerous that children are not allowed out any more. The retreat of pedestrians and cyclists of all ages will continue. As traffic increases, fewer people try to cross the street – one of the reasons why diminishing numbers of people know their neighbours on the other side of the street.

It Will Be More Hostile to Children

Children's freedoms will be further curtailed by parental fears, and the social catalyst of children playing in the street will disappear. In Britain, as recently as 1971, 80 per cent of 7 and 8 year old children got to school on their own unaccompanied by an adult. Now virtually none do, and the Government issues guidance to parents warning that allowing children under the age of 12 out of the house unaccompanied is irresponsible. As the world becomes ever fuller of traffic it becomes increasingly full of strangers; primary schools routinely run "Stranger Danger" campaigns – amplifying parental fears and inculcating paranoia at a tender age. Children become captives of the family chauffeur. The loss of traditional childhood freedoms denies them the experience of mixing independently with their peers and learning to cope without adult supervision, experience essential to the process of socialisation.

It Will Be Fatter and Less Fit

Children with parental chauffeurs no longer acquire the habit of walking or cycling to school, friends or other activities. As functional walking and cycling disappear, we will have less exercise built into daily routines, although this is a trend that appears to be being partially offset by the growing numbers of people who drive to health clubs to run on treadmills. The US Centre for Disease Control and Prevention identified America's dependence on the car as the principal cause of the country's epidemic of obesity, declaring that "decades of uncontrolled suburban sprawl conceived around the motor car have left Americans unable to walk even if they wish to".[2] And the return of infectious diseases like tuberculosis to the developed world is attributed, at least in part, to the growth of international air traffic.

---

[2] Despite the concern about rising levels of obesity in Britain, and the associated campaign against junk food, the real culprit appears to be declining levels of exercise. On average Britons today consume 750 fewer calories per day than 30 years ago – but burn off in exercise 800 fewer calories (House of Commons Select Committee on Health, 2004).

It Will Be Less Culturally Varied

The McCulture will be further advanced. Tom Wolfe captures the phenomenon in *A Man in Full*: "the only way you could tell you were leaving one community and entering another was when the franchises started repeating and you spotted another 7-Eleven, another Wendy's, another Costco, another Home Depot". Tourism becomes an industry. Travel writers urge their readers to rush to spoil the last unspoiled areas on earth, before others beat them to it. The moving pavement that now speeds tourists past the Crown Jewels in the Tower of London to maximise throughput is but one example of the triumph of Fordist efficiency that now characterises mass tourism

It Will Be More Anonymous, Less Trusting and More Paranoid

Fewer people will know their neighbours. Gated communities and Neighbourhood Watch – attempts to recreate some of what used to happen naturally – are symptomatic of the angst of anomie. Even when they live in close physical proximity to each other the mobile wealthy and the immobile poor live in different worlds. The poor are confined by their lack of mobility in prisons with invisible walls. They are continually tempted and taunted – in a way that prisoners confined to cells with opaque walls are not – by the freedom and conspicuous consumption of the affluent. The wealthy can be seen and heard flying overhead, or driving along motorways through the ghetto, or on television, enjoying privileges that remain tantalisingly out of reach. To the wealthy, the poor are often invisible; because of the height and speed at which they travel, the wealthy tend to see the world at a lower level of resolution. A hypermobile world is full of strangers, and people place less trust in strangers than people they know. Lack of trust is conducive to paranoia.

It Will Be More Crime Ridden

The strained relations between haves and have-nots will generate more crime and fear of crime.[3] As with danger on the roads, this phenomenon is not reliably captured by crime statistics. Homes become better defended with stronger doors and locks and alarm systems. People, especially women, retreat from the areas where they feel threatened, especially the streets and public transport, and growing numbers of motorists travel with their doors locked. Policing will become more Orwellian.

Orwellian is the only adjective that can be applied to the vision of the Department of Trade and Industry's Foresight Directorate. The Directorate's Crime Prevention Panel published a consultation document entitled *Just Around the Corner* (DTI, 2000). It surveys the potential for new technology to "create new opportunities for crime and crime prevention". It concludes with two scenarios.

---

[3]   And, since 11 September 2001, more fear of terrorism.

The first, "TECHies" (Teleworking Executives Co-Habiting) is the Directorate's *optimistic* scenario, in which advances in crime-prevention technology outpace advances in crime-promotion technology. It might best be described as *1984* with a *Brave New World* gloss – but which appears oblivious to Huxley's satirical intent. It depicts a world in which *identity theft* is kept in check by all-pervasive surveillance technology, DNA fingerprinting, odour detectors and probabilistic profile matching. The second "socially exclusive" scenario is less cheerful – *1984* without the gloss: most people live in walled estates and don't venture out much because "all public space is potentially hostile". With the rising tide of refugees and the destruction of the World Trade Center by terrorists, the Foresight Directorate's grim vision is acquiring a global reach. Gated communities are being superseded by gated nations.

This high-tech policing, decried by civil libertarians, is an inescapable cost of hypermobility. The alternative is ineffectual policing. If terrorists and criminals avail themselves of modern means of mobility – physical and electronic – and the forces of law and order do not keep pace, the latter will become impotent.

It Will Be Less Democratic

Individuals will have less influence over the decisions that govern their lives. As we spread ourselves ever wider and thinner in our social and economic activities the geographical scope of political authority must expand in order to keep up with the growing size of the problems that require governing. Political authority migrates up the hierarchy from Town Hall to Whitehall, to Brussels and ultimately to completely unaccountable institutions like the World Bank and the World Trade Organization.

Democracy is government by the people. Its purest form (setting aside the plight of women and slaves) is widely held to be Athenian democracy – everyone in the forum had an equal say. Beyond a certain scale this becomes impractical, and the preferred model becomes representative democracy. But as the scale of issues requiring collective management increases still further, representative democracy also breaks down. Either the number of representatives becomes unmanageable and the limits of the Athenian model are reached again – that is the forum becomes overcrowded – or the number of voters per representative reaches a level that renders the individual voter insignificant.

On neither side of the confrontations in Seattle, Prague or Genoa between the advocates of globalisation and disparate groups of protesters could one find institutions that were democratically accountable – Greenpeace and Friends of the Earth are not representative democracies. Trust in these unaccountable institutions diminishes as their "facts" become increasingly difficult to distinguish from spin. In the whole of the genre of science fiction devoted to speculating about futures in which distance has been conquered by science and technology one can find no plausible examples of democracy. The form of government is invariably tyrannical hierarchy. The possibility of an individual voter being of any significance is defeated by scale.

The Governance Agenda for Transport

The trends that are creating the world described above have evolved from combinations of societal trends, in which increasing mobility has played an influential role. They are meeting no effective resistance. On the contrary, they are being encouraged by governments everywhere.

When asked to discuss integrated transport policy development, policy-makers generally claim that transport policy is well integrated with other areas of governance. What they mean is that it is integrated with the planning system. But there are few occasions where the transport elements of this system are well integrated with environmental policy, and none where they are integrated with policies that confront the growing social problems described above.

Transport Governance in Britain

Airport planning in Britain provides a good example. It continues to be based upon the *predict-and-provide* principle – and further vast growth is predicted. Airport planners everywhere reassure each other of the growth potential of their industry by noting that most people in the world have never flown; and the idea that this growth might be constrained by their failure to provide sufficient capacity is, to them, unthinkable.

On the ground, the current UK Government has now abandoned its unconvincing pretence that it wished to reduce the nation's dependence on the car. Gus Macdonald, until recently Britain's Transport Minister, proclaimed the Government's support for increasing it: "If cars become more affordable and more people want to own them, that," he says "is not a problem." He placed the Government firmly in the technical-fix camp – "cleaner engines are the way forward". And John Redwood, a recent transport spokesman for the Conservative Party, not to be outdone in the pursuit of the motorist's vote, urges the construction of more roads to bypass "environmentally sensitive towns villages or beauty spots", forgetting the lesson painfully learned by his Conservative predecessors when in office, that there is a severe shortage of insensitive areas through which to build them.

What would be the principal feature of a policy that sought to *increase* dependence on the car? It would be a package of measures designed to encourage people to move out of town and spread themselves about at densities that were too low to be serviced by public transport. This policy under the previous government met with impressive success; a 1999 study by the Town and Country Planning Association (Breheny, 1999) reports the loss of 500,000 urban jobs and an increase of 1.7 million low-density jobs between 1981 and 1996.

A policy that sought to *reduce* dependence on the car would seek to restrict traffic in the areas where its growth is fastest – not in congested urban areas, where it has already stopped, but in the suburbs and beyond. Private sector consultants are now appearing, offering advice on relocation away from city centres. This free-enterprise equivalent to the old Location of Offices Bureau is a completely unsurprising market response to the additional centrifugal incentives now being devised by the Labour Government in the form of urban road pricing and work place parking charges. Deputy Prime Minister John Prescott insists that he is not anti-car. He, like his Transport Ministers, is happy for more people to own cars but he does, from time-to-time, express the wish that they would leave them in the garage more of the time. He should perhaps replace his road-building programme with a garage-building programme.

When people acquire cars they look for somewhere to drive them and park them, and they rarely find either in Britain's cities. If the nation's car population continues to increase, and the Government's forecasters predict that it will grow substantially, the urban exodus will continue and dependence on the car will increase. Can Britain afford alternatives to the car? Of course. There is no shortage of money. In each of the last four years over £30 billion has been spent on new cars alone.

The Government's enthusiastic promotion of the internet frequently includes the contention that it will help to solve the transport problem by obviating the need for much physical travel. This hope rests upon a decoupling of the trends of electronic and physical mobility for which there is no precedent. Historically, the growth trends of both sorts of mobility have correlated strongly and positively, and today the most physically mobile societies are also the heaviest users of all forms of telecommunications.

Advocates of telecommunications as a part of the solution to present transport problems argue that they will revive and promote human-scale community life by permitting more people to work from home, thereby encouraging them to spend more time close to home, and helping them to get to know their neighbours better. Perhaps. But it presumes that people will be content to lead a shrinking part of their lives in the *real* world which they will experience directly, and a growing part of their lives in *virtual communities* which they will experience electronically. It presumes that people will be content with lives of increasing incongruity of experience – that they will not want to meet and shake hands with the new friends that they meet on the internet; that they will not seek first-hand experience of the different cultures that they experience vicariously electronically; and that they will not wish to have *real* coffee breaks with their fellow workers. It presumes much for which there is, as yet, little encouraging evidence.

I offer a bit of discouraging evidence, albeit anecdotal, from a chance encounter in Vancouver airport while waiting for a flight to London. I got chatting to the fellow sitting next to me who was waiting for a flight to Toronto. He was flying for a game of bridge with someone from Toronto, someone from Scotland and someone from San Francisco. They had met and played bridge on the internet, and now they needed a "real" game. While writing this chapter I listened to a BBC programme on "virtual tourism": without touching fragile environments or cultures it will simulate not only the view but also the noise, smells and even the weather of remote parts of the world which will be spared an invasion by real tourists. The complete lack of irony with which this vision was put forward suggests that its proponents could not have read *Brave New World* – which it mimicked perfectly.

In *Bowling Alone* Robert Putnam documents the rise and decline of civic engagement in American life over a century of increasing physical and electronic mobility. Putnam has amassed an extraordinary range of indicators of "social capital", ranging from membership in Rotary Clubs and bowling leagues to the decline in hitchhiking and participation in parent-teacher associations. Putting all his indicators together, he found that the median peak year for civic engagement in America was 1959 – perhaps an approximate marker of the country's transition from hypo- to hyper-mobility. Putnam favours television – a weak one-way form of electronic mobility – as the principal villain in his story of civic decline. But he also gives an important share of the credit to "sprawl": "this physical fragmentation of our daily lives has had a visible dampening effect on community involvement".

As a sense of involvement dwindles, trust in the institutions that govern our lives also diminishes. Figure 7.2 from a survey by Marris et al. (1996) reveals a remarkable lack of trust in established institutions – trade unions score only 27 per cent, religious organisations 22 per cent and government a miserable 6 per cent. Companies at 9 per cent do little better. Environmental organisations score an impressive 76 per cent, while family and friends, i.e. those likely to have the least expert knowledge about environmental risks, score the highest.

Scientists scored 49 per cent, but a MORI poll cited by Marris et al. (1997) found that approval ratings for scientists were strongly influenced by information about the scientist's employers: top, at 78 per cent, came scientists who worked for environmental NGOs, bottom came government scientists with 32 per cent.[4] The media, directly or indirectly the source of most people's knowledge about environmental risks, score only 15 per cent. Not only do the media inspire little trust, their coverage of environmental issues is widely ignored. Hypermobility fosters political apathy which in turn generates a disinterest in debates about issues that are beyond the influence of the individual citizen. At the height of the Brent Spar controversy, an issue which received enormous media coverage, only 59 per cent of those questioned about Brent Spar were aware of the incident. Only the doctor, amongst traditional institutions, retained a respectable level of trust.

Transport Governance in Europe

There has been an impressive growth in recent years in an interest in governance, both on the part of governments and the corporate world. Typing the word into Google, at the time of writing, yields 7.6 million hits; "corporate governance" yields three million. The European Commission has published a White Paper on the subject (European Commission, 2001) in which it presents its own concept of governance as referring to:

---

[4] The widespread suspicion of conspiracy between government and industry, and the mistrust of science sponsored by either, was highlighted by Colin Blakemore, president of the British Association for the Advancement of Science, in his call for Britain's Minister of Science to be detached from the Department of Trade and Industry and given an independent position in the cabinet (*The Times*, 3 September 1998).

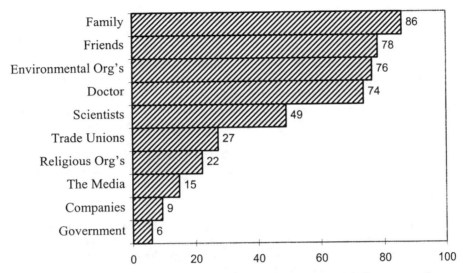

Combined and weighted results from four samples: Chamber of Commerce, Scouts, Greenhouse (a green environmental organisation), and a general sample.

Figure 7.2 Percentage of respondents who said they would "often" or "always" trust institution X to "tell them the truth about risks"

the rules, processes and behaviour that affect the way in which powers are exercised at European level, particularly as regards openness, participation, accountability, effectiveness and coherence.

It observes that:

Today, political leaders throughout Europe are facing a real paradox. On the one hand, Europeans want them to find solutions to the major problems confronting our societies.

On the other hand, people increasingly distrust institutions and politics or are simply not interested in them.

And:

despite its achievements, many Europeans feel alienated from the Union's work.

The president of the World Bank has also taken an interest:

Corporate governance is about promoting corporate fairness, transparency and accountability (*Financial Times*, 1999).

Figure 7.2, and other Mori polls,[5] suggest that the growth in the number of high-level expressions of concern about openness, participation, accountability, corporate fairness and trust are negatively correlated with trends in popular belief in the effective existence of such phenomena. High-level responses to these concerns are largely platitudinous. In 2002, the European Commission published a report entitled *Making globalization work for everyone: The European Union and world trade* (European Commission, 2002). Consider the likely impact on the disengaged, alienated, non-voting EU citizen of this passage from the report:

> As the world economy globalises, the WTO is the most legitimate forum for removing obstacles to trade, creating and enforcing global rules and making them compatible with rules drawn up by other multilateral bodies. The aims of the EU's work in the WTO effort are:
>
> - to open up markets in goods, services and investment in accordance with clear rules and following a timetable that enables all countries to implement them;
> - *to make the WTO more open, accountable and effective by engaging in discussion with other groups and organizations* [my italics];
> - to bring developing countries fully into the WTO's decision-taking processes, helping them to integrate with the world economy.

Worldwide

As international travel becomes faster, cheaper and easier for the wealthy it becomes more difficult, bureaucratically, for the poor. Wealthy countries previously protected by distance from mass invasion by the indigent are increasingly resorting to restrictive prohibition and force. Barriers – in the form of stringent visa requirements, difficult-to-obtain work permits, and obstructive immigration requirements – are being raised to contain the numbers who seek to take advantage of the mobility afforded by technology. The "huddled masses" who used to be welcomed to America by the Statue of Liberty are now dubbed economic migrants and denied entry to protect the living standards of those who got there earlier.

Throughout history most people in most places have led pedestrian lives. Their settlement patterns and travelling have, as a consequence, been tightly constrained. Such vehicular transport as existed was powered by humans, animals or wind. The rich had more mobility than the poor, but no one had very much. Mythologies abounding in advance technologies – flying carpets, seven-league boots, winged chariots and the like – attested to a pervasive desire for more, but in technologically unimaginative ages most people were resigned to this remaining the prerogative of the gods. Indeed the legend of Icarus suggests that the very idea of mere mortals attaining such means of travel was an impious one.

At a time that roughly coincides with the beginning of the industrial revolution in England there began a period of remarkable reductions in the cost of transport and even more remarkable increases in its speed and comfort, and in the numbers

---

[5]   See http://www.mori.com/ and O'Riordan et al. (1997).

who made use of it. The achievements of the gods have been surpassed. Concorde could fly faster than Apollo's flaming chariot, and advances in telecommunications have created a capacity for exchanging information that far exceeds anything ever attributed to Mercury. The transport and communications history of this period is almost invariably told as a story of progress following in the train of technological advance. And any problems associated with this progress have been seen as "side effects" treatable by yet more technology. *Hypo*mobility was bad. More mobility is good. *Hyper*mobility? Might it be possible to have too much of this good thing? This is not a question that has been seriously addressed by historians of transport, or planners and politicians concerned with its future. Even to raise the question risks labelling oneself an anti-democrat – an enemy of freedom and choice.

This risk can be reduced if one is careful about the way one puts the question. Everywhere in the world the "transport problem" can be usefully put in three parts, in the form of three opinion polls.

- The first question is asked frequently. *Would you like a car, unlimited air miles and Bill Gates' level of access to all the electronic modes of travel?* With minor variations this simple question is routinely put by opinion pollsters, and worldwide the answer is overwhelmingly YES! This is the implicit opinion poll that still sets the political agenda for transport planning almost everywhere. In answering, people imagine the world as it now is, but with themselves gaining access to the greater range in opportunities in life that they see the wealthy enjoying. Most politicians believe it would be political suicide to resist such aspirations. It would be manifestly unfair, they often add, for those who already enjoy a high level of mobility to pull the ladder up behind them. The ladder metaphor, originally invoked by Anthony Crosland in a Fabian lecture almost thirty years ago, makes a strong moral point that translates readily into a political imperative.
- But, there is a second question which is never asked. *Would you like to live in the sort of world that would result if everyone's wish were granted?* Assistance with the answer might be given by rephrasing the question – *would you like to live in a dangerous, ugly, bleak, crime-ridden, alienated, anonymous, undemocratic, socially polarised, fume-filled greenhouse threatened by terrorism without precedent?* The "fume-filled greenhouse" is optional; I strongly suspect that technological improvements will not keep up with traffic growth, and that the physical environment will deteriorate as mobility levels rise; but confining the question to the *social* consequences described above should be sufficient to elicit the answer NO. This opinion poll asks, in effect, do you want the consequences of "business as usual"? As these consequences become better, and more widely understood, increasing numbers of people are clear that they would not want them. But the political response has been disappointing. The best that even progressive Denmark or the Netherlands have achieved so far is a response that slows the rate of growth in road traffic in urban areas, does little to slow the growth of traffic in the suburbs and rural areas, and does virtually nothing to arrest the far more rapid increase in air travel.

Crosland's ladder has become an extension ladder that is still extending. Britain's Department of Transport describes the continued growth of traffic as "inevitable" – cheerfully ignoring the fact that those on the bottom rungs of this ladder are being pushed deeper into the mire of social exclusion, and are manifesting their resentment in ever more terrifying ways. The political difficulty seems to be that the problem, when posed in the form of Opinion Poll 2, implies the need for a grim, grey, virtuous self-denial in order to save the planet. This is not a platform on which many politicians are enthusiastic to campaign.

But there is a third, more cheerful question – the inverse of the second question. *Would you like to live in a cleaner, safer, healthier, friendlier, more beautiful, more democratic, sustainable world in which you know your neighbours and it is safe for your children to play in the street?* If these rewards could be assembled in a convincing and affordable package most people could be expected to vote for them – especially if the consequences spelled out in Opinion Poll 2 were seen as the alternative.

For most people in the world, realisation of the aspirations encapsulated in the first opinion poll is a vanishing possibility. But so long as its pursuit continues to be the principal objective of transport planners and policy-makers, the achievement of the bleak scenario set out in the second question becomes more likely. However, contrary to the assertion of Britain's Transport Minister, the rising tide of traffic which is bringing it about is not inevitable. This traffic tide is not an irresistible force of nature, like the oceanic tide, to which we can but adapt. It is the consequence of myriad human decisions large and small – of decisions by governments, about taxes and subsidies, about land use planning, about road and airport building, and of individual responses to these decisions. It is driven by a deeply-rooted, reality-denying, linear view of progress.

The first question is equivalent to asking a glutton if he would like unlimited quantities of his favourite foods and drinks. The answer is predictable. The second question confronts the glutton with the consequences of unconstrained indulgence. There are expensive, high-tech solutions to some of these consequences – liposuction, Olestra (the fat that slips straight through), and by-pass surgery (here the analogy gets particularly close). But eating less and walking or cycling to work are likely to be more effective, save money, and produce a greater sense of well-being and self-worth.

Achieving the society encapsulated in Opinion Poll 3, which appears impossible to most politicians, is *in principle* quite straightforward. It simply requires a reordering of priorities. Instead of continuing to sacrifice the physical and social environment for more mobility, it requires fostering the local at the expense of the remote, and foregoing some of the benefits of mobility to protect and enhance what we value in nature and our relations with friends and neighbours. To question the benefits of hypermobility is not to deny freedom and choice. It is to ask people what it is that they really, *really* want, and to confront them with the fact that their choices have consequences beyond the primary objects of their desires. Collective self-discipline is the wise exercise of freedom and choice.

Simple – *in principle*. In practice, the longer the linear logic connecting increasing mobility to progress continues to guide transport and communications policy, the more remote this possibility becomes. Bottom-up "governance" – as distinct from top-down government – is not possible in a fast-moving, anonymous, low-trust, paranoid hypermobile world.

Acknowledgement

This chapter is a revised version of a lecture given to the Royal Society for the Encouragement of Arts, Manufacture and Commerce (RSA) 'The Social Consequence of Hypermobility', 21 November 2001 www.rsa.org.uk/acrobat/hypermobility.pdf. This lecture drew in turn from a report for the OECD *The Social Implications of Hypermobility* www.olis.oecd.org/olis/1999doc.nsf/LinkTo/ENV-EPOC-PPC-T(99)3-FINAL-REV1 – starting page 95).

References

Breheny, M. (ed.) (1999), *The People: Where Will They Work?*, London: Town and Country Planning Association.

DTI (2000), *Just Around the Corner*, Consultation document issued by Foresight Directorate's Crime Prevention Panel (http://www.foresight.gov.uk/).

European Commission (2001), *European Governance: A White Paper*, 25.7.2001, COM(2001) 428 final, Brussels: European Commission.

European Commission (2002), *Making globalization work for everyone: The European Union and world trade*, Brussels: European Commission.

Financial Times (1999), 'Speak up for dialogue', 21 June 1999, London.

House of Commons Select Committee on Health (2004), *Obesity*, HC 23-III, London: The Stationery Office. See Written Evidence, Appendix 18, Memorandum by the Royal College of General Practitioners (OB 31) April 2003 www.publications.parliament.uk/pa/cm200304/cmselect/cmhealth/23/23we19.htm.

Huxley, A. (1955), *Brave New World*, Harmondsworth: Penguin.

Marris, C., Langford, I. and O'Riordan, T. (1996), *Integrating sociological and psychological approaches to public perceptions of environmental risks: detailed results from a questionnaire survey*. CSERGE Working Paper GEC 96-07, University of East Anglia.

O'Riordan, T., Marris, C. and Langford, I. (1997), 'Images of science underlying public perceptions of risk', *Science Policy and Risk*, London: Royal Society.

Orwell, G. (1974), *1984*, London: Secker and Warburg.

Putnam, R. (2001), *Bowling Alone: The Collapse and Revival of American Community*, London: Simon & Schuster.

The Times (1998), 'Call for new ministry to deal with scientific challenges', 3 September 1998, London.

Wolfe, T. (1999), *A Man in Full*, New York: Picador.

Chapter 8

Environmental Policy Integration for Sustainable Technologies: Rationale and Practical Experiences at EU Level

Julia Hertin and Frans Berkhout

There are many different ways in which policy approaches to science and technology, risk and the environment could be better integrated or – to put it more modestly – co-ordinated and aligned.[1] For example, integration may be needed across different vertical levels of policy-making, amongst different governmental and non-governmental actors and throughout the process of technological innovation. One form of integration has received particular attention in recent years – both in the policy literature and the political practice: the integration between environmental policy and key sectoral policy areas such as transport, planning, agriculture, energy and so on.

This chapter explores this issue of environmental policy integration and its relevance for a governance approach oriented towards the development of sustainable technologies. It starts by asking why the traditional model of sectoral environmental policy has come under increasing attack, especially from an innovation perspective. It then aims to make the concept of environmental policy integration more tangible by describing its key elements and how it is thought to contribute to better governance. The final part turns to practical experiences with environmental policy integration in EU policy-making. Providing an analysis of recent initiatives such as the EU Cardiff process, the Sustainable Development Strategy and the Impact Assessment procedure, we explore the opportunities for, and barriers to, environmental policy integration.

---

[1] Parts of this chapter draw on two previous publications: Hertin, J. and F. Berkhout (2002), 'Practical experiences on policy integration and recommendations for future initiatives on EU and national level', Paper for the 3rd Blueprint workshop: *Instruments for Integrating Environmental and Innovation Policy*. 26-27 September 2002, Brussels; Hertin, J. and F. Berkhout (2003), 'Analysing institutional strategies for environmental policy integration: The case of EU enterprise policy', *Journal of Environmental Policy and Planning* 5(1): 39-56.

The Rationale for Environmental Policy Integration

When the environment was beginning to be established as an independent policy field in the early 1970s, most industrialised countries set up institutions specifically dedicated to the protection of the natural environment, for example ministries, agencies, reporting mechanisms, framework laws and expert advisory committees (Jörgens, 1996). This pattern of institutionalisation followed the Weberian model of rational and effective administration. It is based on functionally differentiated organisation, a principle that is virtually a universal characteristic of public administration. It involves the allocation of manageable areas of policy to specialised units. These can develop the necessary expertise and build up policy networks with stakeholders inside and outside government, thereby enabling the organisation to respond to the complexity associated with specific policy areas.

This model manifested itself in a specific environmental policy approach. Environmental ministries were given responsibility for mitigating and dealing with the effects of pollution. They approached ecological problems technocratically: treating different aspects of pollution separately, applying standard routines of pollution abatement, combating symptoms rather than tackling sources (Jänicke, 1990). This segregated institutionalisation of the environment neglects the source of pollution problems which occur as the result of a wide range of sectoral activities such as industrial production, transport, and agriculture. Sectoral policy initiatives covering these fields, however, are commonly formulated with little or no regard to the environment. This might be due to a lack of capacity, but also reflects more deeply-rooted belief systems, professional cultures, sources of information, and institutional arrangements (cf. Weale, 1992). The perceived role and mission of sectoral departments has traditionally not included environmental concerns. For example, the overriding concern of energy departments has conventionally been to supply cheap and secure energy, rather than to minimise air pollution or to prevent climate change. In addition, there has been a belief that better decisions would arise from institutional arrangements that established a competitive relationship between promoting and regulating departments. A proper contest of ideas is seen as more likely to deliver clarity and to improve the rationale and evidence base for a given policy action.

In principle, the problem of a "misfit" between the structure of environmental problems and the structure of the administrations created to address them could be overcome through conscious co-ordination efforts. But co-ordination between administrations that may have antagonistic interests has proven difficult, and does not encourage the co-operative relationships that may be necessary to encourage flexibility and learning (Hertin and Berkhout, 2003). This partly reflects tensions between the respective political constituencies, but is also due to the fact that environmental policy initiatives often "infringe" on the responsibilities of sectoral administrations. As a consequence, status, power and resources are often at stake in environmental disputes, placing an additional strain on inter-departmental relationships (Weale, 1992). The combination of an administrative "misfit" and antagonistic relationships leads to patterns of communication and decision-making which are at odds with an innovation-oriented environmental policy. In particular, we identify lack of incentives for innovation, a bias against integrated responses, and the problem of an unstable context for innovation to take place.

Lack of incentives for innovation   Environmental departments rarely prevail in inter-departmental bargaining processes. Historically they have a lower status than sectoral departments such as transport, energy and agriculture, mainly because their interests are asymmetric (Wilson, 1980). While polluter interests tend to be driven by strong economic motives and represented by well-resourced advocacy groups, environmental interests are more dispersed and possess less direct political leverage. As a result, it is a common experience that the ambitions of environmental proposals are considerably reduced during intra-governmental negotiations. A policy that does not sufficiently depart from the status quo is unable to provide both the economic incentives and the political signal needed to stimulate the innovation and diffusion of new technologies (Wallace, 1995; Blazejczak et al., 1999).

Bias against integrated technological responses   In order to minimise "interference", sectoral departments have an interest in reducing the involvement of environmental administrations in the process of policy formulation. This tends to delay inputs from environmental interests. When the environmental ministry voices its concerns in the later stages of sectoral policy design, the opportunities to embed environmental safeguards are limited. Inter-departmental conflicts are resolved at this stage through compromises that satisfy the interests of both agencies but may be highly inefficient. Either contradictory policies are implemented in parallel or costly end-of-pipe technologies are added to a controversial project to make it acceptable for environmental departments (Jänicke, 1990). The relationship between intra-governmental relationships and policy outcomes can be self-reinforcing: "For government, the self preservation instincts of those institutions that have been established to administer this form of policy [additive pollution control] ensure that this approach continues" (Gouldson and Murphy, 1996: 12).

Unstable innovation context   The processes of bargaining associated with conflictual intra-governmental policy-making lead to mixed political signals and uncertainty about policy direction. This creates political conditions tending to discourage environmental innovation, i.e. a reliable, stable, knowledge-based and long-term policy based on consensual decision-making (Blazejczak et al., 1999). The resulting absence of a stable perspective for the emergence of markets acts as a barrier to investment in research and development, and slows the pace of innovation (Wallace, 1995).

In short, it can be argued that departmentalised administrative structures create a bias towards policies that are unable to exploit the environmental potentials of modern technology. This line of argument has been strengthened by the emphasis recent literature has placed on understanding environmental innovation from a systems perspective (Kemp and Rotmans, 2001; Berkhout and Gouldson, 2003). Here, successful protection of the environment is seen as an integral part of long-term policy strategy shifts in key socio-technical systems supporting the provision of food, transport, and energy. This is based on the recognition that environmental sustainability requires large-scale transitions of socio-technical systems that are unlikely to be delivered by incremental changes that are predominant in the systemic model of innovation (cf. Freeman, 1996). A transition to a renewable energy system, for example, is likely to require changes to many aspects of the electricity system (generation, transmission, distribution), other power generation and heating systems (industrial and households), many aspects of the transport system (modes of transport, transport fuel systems, propulsion systems) and probably radical energy-saving measures in all sectors. The transition will only be possible if it is supported – or at least not obstructed – by a range of industrial policies, and by science and technology policy, fiscal policy, infrastructure policy, spatial planning and so on.

Key Elements of Environmental Policy Integration

Environmental policy integration can generally be taken to describe the inclusion of environmental concerns in processes and decisions of public policy-making which are predominantly charged with issues other than the environment. However, it often remains ambiguous what exactly this integration entails: does it describe a procedural requirement or is it an "autonomous normative principle" (Nollkaemper, 2002)? What precisely should be integrated: policy objectives, decision-making structures, knowledge and capabilities, or policy instruments? Does it involve a changed balance of power between sectoral and environmental administrations, or is integration a question of expertise and organisational routines? Because environmental policy integration is a multi-faceted concept, it is useful to break it down into a number of key elements.

Sectoral Agenda Setting

Environmental policy integration requires first of all that environmental concerns are put on the political agenda of sectoral administrations. A shift from sectoral to an integrated policy can only occur if the administration is aware of unintended environmental effects of sectoral initiatives, recognises them as relevant for its own strategy and takes steps to improve environmental outcomes. Specific examples of agenda setting could be, for example: inclusion of environmental objectives into the overall mission statement of a department; presence of environmental mechanisms in standard operating procedures; and inclusion of environmental provisions in sectoral policy initiatives and outputs.

Horizontal Communication

Successful integration does not make environmental administration redundant. Relying exclusively on environmental expertise in the departments of transport, agriculture, energy and so on would involve building up inefficient duplicate structures and neglecting the generic elements of environmental policy. Therefore, integrated environmental policy would also have to be based on well-functioning horizontal communication between administrative units with different responsibilities. The interaction between sectoral and environmental administrations can take different forms: formalised inter-departmental consultation processes, issue-specific joint working groups, frequent ad hoc meetings and informal discussions, etc.

Sectoral Capacity Building

Horizontal communication will only lead to integrated policies if the sector administration possesses the capacity to appropriate, process and implement environmental knowledge (Schout and Jordan, 2005). Assessing the potential ecological effects of policies is a difficult, contentious and uncertain task, especially where complex and wide-ranging changes to a system of rules (e.g. changes to the land use planning system) are planned. Specific expertise is also needed to recognise "win-win" opportunities and to exploit them through the formulation of innovation-oriented policies. The question of how environmental innovation capacity is best institutionalised will depend on the specific context of the sector. Institutional forms could include specialised environment units, a system of officials charged with environmental responsibilities (for example, Green Ministers), internal information services, training, and specific budgets for external consultancy.

Policy Learning

Environmental policy integration is aimed at a more fundamental process of policy learning, understood as a change of values and belief systems over time (Sabatier and Jenkins-Smith, 1988). Public policies are based on a certain framing of a problem, assumptions about causal relationships, and so on (Pressman and Wildavsky, 1979). If sectoral (and environmental) departments "learned" that environmental concerns can be reconciled with other policy objectives, damaging turf disputes would be avoided and strategies to improve the environmental efficiency of goods and services could be promoted. While it might not always be possible to distinguish a change in political rhetoric from genuine changes of attitude, there are developments which can be taken as indications of policy learning (e.g. changed framing of environmental issues in policy documents, changed attitudes in inter-departmental negotiations, different outcome measures, changed appraisal and incentive structures for administrators).

The Challenge of Putting Policy Integration into Practice

How can environmental policy integration be implemented in practice? One can distinguish at least four different types of institutional mechanisms that governments have used to promote integration. It should be noted that these represent ideal-types that hide a great deal of complexity in practice.

Integrated departments Merging the environmental department or ministry with an environmentally sensitive sectoral ministry is, at first sight, an obvious integration strategy. In the UK, for example, environmental protection, agriculture, food and rural affairs were integrated into one department following the 2001 General Election. This was as a response to a series of perceived failures in the previous Ministry of Agriculture, some of which had environmental implications (nature conservation in agriculture, for instance). The aim of integrated ministries is to improve communication between formerly separated departments to develop integrated policy responses at a lower hierarchical level. Merging departments can also be expected to raise environmental issues on the agenda of the sectoral parts of the administration. Building new environmental capacity is not an inherent element of this strategy, which focuses on bringing the different perspectives closer together. Policy learning can occur within integrated departments although environmental concerns tend to be imposed on the non-environmental division of a new department. An obvious limitation of this approach is that it can only include a limited number of sectors.

Communication mechanisms In recent years, many EU countries have strengthened mechanisms for horizontal communication and co-operation between environmental and other sectors, e.g. through inter-ministerial working groups and committees, consultation procedures, or environmental "correspondents" in sector ministries. Horizontal co-operation mechanisms will usually make the policy sector more aware of environmental issues (agenda setting), but they are unlikely to initiate a more fundamental learning process. Most co-operation mechanisms also do not appear to build up capacity but aim to integrate the expertise of the environment sector into other policy areas through better horizontal communication. Better linkage of expertise and policy communities of interest may indeed be a way of reducing the need for sector-based capacity building. The effectiveness of these formal procedures, however, depends on the willingness of the co-operating institutions and individuals to use them constructively, and the resources and incentives that are made available.

Central strategy Integration can also be promoted through a political rather than an administrative strategy. This provides political leadership and an overarching framework for an integrated governmental approach. It typically consists of a high-level mandate, an environmental or sustainable development strategy, and quantitative targets. The Dutch National Environmental Policy Plans and the UK Sustainable Development Strategy are examples of this approach. This approach provides, first and foremost, the political momentum for promoting environmental issues on the policy agenda across all government departments. If the strategy is developed collaboratively, it will promote communication between the environmental and sectoral ministries. It does, however, not aim to build up environmental expertise in non-environmental departments because the focus is on political engagement, rather than on competence. On its own, it is also unlikely to change fundamentally the perception of environmental issues by sector ministries.

Sectoral integration strategies   A more decentralised approach is to build integration strategies into each policy sector. This can be defined as a process through which non-environmental policy sectors assess the environmental implications of their decision-making, set out a strategy to reduce negative and enhance positive environmental effects, and evaluate the success of the process. This approach underlies the Cardiff process (see below). Sectoral integration directly promotes agenda setting and policy learning as sectors are requested to set themselves environmental objectives and to incorporate these into their policy strategy. This process is expected to lead to the identification of new win-win options and integrated solutions. Developing a sectoral environmental strategy also adds to the capacity of a sectoral department because it typically involves tasks such as assessing the environmental impacts of the sector, defining key environmental issues, and setting targets. Inter-departmental communication might occur during the development of sectoral integration strategies, but the approach relies on a sectoral focus, rather than enhanced co-operation.

These strategies should not be seen as mutually exclusive. Governments commonly apply different strategies in parallel (Jordan and Lenschow, 2000).

Practical Experiences in EU Policy-Making

While the theoretical case for environmental policy integration is compelling, putting the principle into practice has proven difficult. Over much of the last two decades, the principle of integration has mostly served as a rhetorical reference point. With the exception of a few front-runner states, such as Sweden and the Netherlands, there has been no real effort to overcome the departmentalised character of environmental policy in the European Commission or at the member state level. "Integration" was mainly understood as the integrated management of pollution across different media (air, water, and land) within environmental policy. This is reflected in the integrated pollution prevention and control directive (1996). However, since the early 1990s environmental policy integration has moved gradually up political agendas, as the weaknesses of existing approaches came to be more widely recognised. The main thrust of this trend has come from the EU level, with the Cardiff process, the Sustainable Development Strategy and the recent introduction of Impact Assessment marking particular milestones.

Environmental policy integration constituted one of the objectives in the Third Environmental Action Programme (1982-1986). The Single European Act (1987) formally incorporated the integration principle into the EEC Treaty, demanding that "environmental protection requirements shall be a component of the Community's other policies" (Article 130r). But it was the Amsterdam Treaty (1997) that gave environmental policy integration its current political significance, by introducing it into Part One of the EC Treaty, which established the principles of the Community:

> Environmental protection requirements must be integrated into the definition and implementation of the Community policies and activities referred to in Article 3 [listing of all Community activities], in particular with a view to promoting sustainable development (Art. 6).

Integration was also at the heart of the Fifth Environmental Action Programme (1992-2000) which marked a shift from a reactive and symptom-oriented approach towards a more comprehensive policy strategy aiming to address the causes of environmental degradation.

This emphasis on the integration principle in key policy documents contrasts with a lack of progress in joining-up EU policy-making. EU institutions have traditionally been organised in a hierarchical and segmented fashion (Lebessis and Paterson, 1998). The different sectoral formations of the Council work largely independently, and co-ordinating mechanisms are weak. In the case of the European Commission it appears that inconsistency is not so much due to the lack of communication as an outcome of the structural alignment of interests. A variety of generic co-ordination mechanisms are used, for example inter-service consultation, task forces, corresponding units, and incentives for staff rotation. Instead, the relationship between DG Environment and sectoral Directorate-Generals such as DG Enterprise and DG Agriculture is characterised by deeply entrenched conflicts of interests, cultural differences and controversies which prevent the re-alignment of policy objectives and strategies. As in most national governments, the environmental administration lacks political influence with regard to the larger and more powerful sectoral policy sectors such as internal market and agriculture.

Communication Mechanisms

Soon after the adoption of the Fifth Environmental Action Programme, the Commission agreed on a package of institutional reforms designed to tackle the recognised problem that sectoral policies frequently worked against environmental objectives and commitments. It had five main elements (cf Wilkinson, 1997; Kraack et al., 2001):

1. A system of Green Ministers (so-called Integration Correspondents) was introduced. Their mission was to ensure that policy proposals take account of the environment and sustainability.

2. An Integration Unit was set up in DG Environment[2] charged with monitoring progress towards environmental policy integration throughout the Commission, providing advice and training on environmental appraisal and serving as a point of contact for Integration Correspondents in sectoral DGs.
3. Proposed new legislation with particular relevance (marked with a Green Star) to the environment had to go through a process of environmental appraisal.
4. All DGs were asked to provide an annual evaluation of their environmental performance.
5. A Code of Conduct was introduced to promote better internal environmental management and responsibility (procurement, energy saving, etc.).

Observers agree that this package of horizontal co-operation mechanisms has not delivered substantial progress towards environmental policy integration. Wilkinson (1997) and others (Kraack et al., 2001) have found that the mechanisms have not been fully implemented. For example, the Environmental Correspondents have met only infrequently and do not always play the anticipated role of "guardian of the environment". Instead, they appear to have viewed their mission as fending off requests from DG Environment. Similarly, the internal reporting that was due to be carried out under the Green Star system suffered from a lack of guidance and resources, as well as being very unpopular. As a result, the system was not taken seriously and not fully implemented. Despite being a close observer of European environmental policy, Wilkinson found "...no evidence that any such environmental appraisals have been undertaken" and that "...no references to such appraisals have been included in any Commission document" (1997: 163). An official from DG Enterprise cited as a main reason for this failure of the Green Star system the perception that sectoral DGs resented the system because they felt "controlled" by DG Environment.

Although a full and rigorous assessment of the workings of these mechanisms has not been published, it appears that their impacts have been modest. They may have led to a certain level of negative co-ordination, where the intervention of DG Environment delayed, averted or modified initiatives of high environmental concern. But, they have not contributed to the development of the shared positive vision in each policy sector needed as a focal point for a coherent sustainability strategy that emphasises learning and innovation. The communication mechanisms have also done little to address the problem of the unequal balance of power between sectoral and environmental interests. This was recognised in a working paper published by the European Commission's Forward Studies Unit:

> ...despite the clear need for the integration of policies and despite the increase in the use of tools designed to achieve it, there is evidence from the environment sector, for example, to suggest that there is frequently a tendency towards minimising the possible influence of these tools. It appears that formal processes aimed at policy integration can

---

[2] Then called DG XI. For consistency, we refer to DG Environment throughout the chapter.

become in practice little more than opportunities for representatives of different services to recite fixed positions (Lebessis and Paterson, 1998: 16).

The Cardiff Process

Another important strand in the EU strategy for integrated environmental policy has been the so-called Cardiff process. Its name stems from the Cardiff European Council (1998) which started off the process by requesting key sectoral Councils to "establish their own integration strategies within their respective policy areas" (Presidency Conclusions). The structure of the Cardiff process evolved through an interactive process between the European Commission, successive European Councils and the Council of Ministers. The European Commission played an important part in developing the overall structure of the process with its Communication "Partnership for Integration" (CEC, 1998). The document, largely written within DG Environment, develops practical steps towards integration, proposing environmental assessments for key policy proposals, strategies for action in key policy sectors, as well as indicators and monitoring mechanisms. It acknowledges that "too often in the past attempts to establish horizontal principles have resulted in bureaucratic and mechanistic requirements which have failed to deliver" (p.6). The European Council, although supporting the process during the British, Austrian and Finnish presidencies, played a minor role, largely endorsing the proposals of the Commission. Council formations such as transport, energy, agriculture, industry and others were expected to develop a sectoral environmental strategy, an action plan and mechanisms to monitor progress and to present these at the Gothenburg Council in June 2001.

Different possible explanations are put forward as to why this initiative was adopted in this form and at this time. Some interpret it as a genuine attempt of "green" member states to give practical relevance to the Article 6 commitment. Others are more suspicious. Kraack et al. (2001) note the enthusiasm with which Britain has promoted the Cardiff process and suggest that this may have been a conscious attempt to shift the balance away from the unloved legislative instruments that had dominated EU environmental policy since the 1970s, and towards a greater emphasis on soft (and perhaps less effective) steering mechanisms.

Since the Cardiff European Council, all key Council formations have published either a report on environmental policy integration or a strategy. Early evaluations (Fergusson et al., 2001; Kraemer, 2001; Schepelmann, 2000) have shown that, although the task was essentially the same for every sector, the processes and approaches are heterogeneous and quality varies. While most sectors focused on environmental integration, some chose to include all three dimensions of sustainable development (economic, social and environmental). The degree of institutionalisation of the process (e.g. setting up expert groups), the contribution of the Commission and the involvement of other institutions also varied. Processes are generally not transparent and little consultation has taken place with non-governmental stakeholders. In most sectors, the process seemed to develop in a relatively unstructured and ad hoc way. Officials involved in the Cardiff process expressed uncertainty about what the final outputs and outcomes of the strategy would be.

Overall, the reluctance of many policy sectors to engage with environmental agendas finds expression in the Cardiff process documents. None of the strategy documents contains all the elements expected of a strategy. Problem formulation often remains vague and cursory, and a thorough assessment of the environmental effects of the sector is missing from most strategies. Although Councils have stated aims and principles with regard to environmental policy integration, quantitative targets have not been formulated and real commitments are lacking (EEB, 2001). Most integration reports and strategies place an emphasis on existing policies and those that are currently being developed. They rarely propose or discuss new policy initiatives. But not only is progress towards the formal requirements of the Cardiff process patchy, there is also little evidence to suggest that it has brought the EU closer to resolving key programmatic contradictions.

The Gothenburg European Council (2001) made it clear that there was still further to go in achieving integration and that it expected the European Council formations to "finalise and further develop" their sector strategies (Presidency Conclusions). On the positive side, it can be noted that, by forcing the individual policy sectors to formulate their understanding of "sustainable development", the Cardiff process has made certain lines of conflict more transparent. It has also contributed to a significant amount of environmental agenda setting and capacity building in the sectoral policy institutions. In the more advanced sectors such as energy and transport the Cardiff process has probably helped to spell out elements of a positive vision of sustainability.

Sustainable Development Strategy

At the Lisbon European Council in March 2000, the EU set itself an ambitious agenda of structural reform aiming to make Europe's economy more dynamic and competitive while safeguarding social inclusiveness. It coined the much-cited phrase of aspiring to become "the most competitive and dynamic knowledge-based economy in the world, capable of sustainable economic growth with more and better jobs and greater social cohesion". Despite the mention of the word "sustainable", the environment did not play a significant part in this initial conception of the Lisbon Strategy. Only the Gothenburg European Council the following year added an environmental dimension to this key strategic policy initiative of the EU.

Rather than being conceived as a set strategy laid down in a single text, the EU Sustainable Development Strategy (SDS) constitutes an evolving process recorded in a suite of documents of varying status. Three key documents were published in 2001: a proposal for a SDS drafted by the European Commission; the Gothenburg Presidency Conclusions (the Commission proposal was not formally adopted but many of its key elements were included in the Presidency Conclusions); and a consultation document prepared by the Commission. The adoption of the SDS was not only a response to pressure from environmental organisations that had been requesting a strategy for many years. Most European governments had already adopted national sustainability strategies or environmental policy plans – especially following the UN Conference on Environment and Development in Rio in 1992. Moreover, the Sustainable Development Strategy was the key contribution of the EU to the 10-year follow up to the Rio Summit in Johannesburg, 2002. An additional statement, covering EU external policies, was adopted by the Commission in February 2002. This added a global dimension to the strategy ("Towards a global partnership for Sustainable Development"). The process of developing and monitoring the SDS includes an annual review of progress at the spring European Council and the extension of the existing indicators to monitor the impact of the Lisbon strategy through environmental indicators.

Interestingly, the SDS puts as much emphasis on processes of governance as it does on substance. The first part of both the Commission proposal and the Council Conclusions emphasise the need for a new policy approach. This focus on process is not surprising given that there is no lack of environmental objectives (although often qualitative ones) and policy initiatives. The fundamental weakness in EU policy-making appears to be the lack of integration and co-ordination, rather than a lack of ambition:

> At all stages of the Community legislative process, policy proposals in individual sectors are developed and discussed without paying sufficient attention to the linkages between different policy areas. The way the Commission, Council and Parliament are organized reinforces this narrow, sectoral approach. All three institutions should consider what steps they can take to overcome this weakness (CEC, 2001b: 14).

Five requirements of governance for sustainable development are set out:

1. using market-based instruments to adequately reflect external environmental costs and provide incentives for consumers and producers;
2. improving policy co-ordination at the level of the Member States through national sustainable development strategies and national stakeholder consultation;
3. improving policy coherence in the EU through better policy co-ordination in the Council and the introduction of sustainability impact assessment for all major policy proposals;
4. annually reviewing the implementation of the SDS on the basis of headline indicators, the Sixth Environmental Action Programme and the Cardiff sectoral strategies;
5. continuing to develop and implement the Cardiff sectoral integration strategies.

On substance, the SDS includes a strongly-worded commitment about environmental sustainability which emphasises the social and economic opportunities associated with it: "Clear and stable objectives for sustainable development will present significant economic opportunities. This has the potential to unleash a new wave of technological innovation and investment, generating growth and employment" (CEC, 2001b: p4). The Gothenburg Presidency Conclusions set or re-affirm a number of commitments in four areas of environmental concern: climate change (a 22 per cent renewable electricity target by 2010); transport (decoupling of transport growth and GDP growth and a shift from road to rail, water and public passenger transport); public health (the aim to have the proposed chemicals policy in place by 2004); and natural resource management (increasing the emphasis of the Common Agricultural Policy on health and environment, adapting the EU fishing effort to the level of available resources, and halting general biodiversity decline).

Environmental observers have interpreted the SDS process and the link to the Lisbon agenda as an important step towards environmental policy integration. On the other hand, it has also been noted that the political will to give greater emphasis to environmental objectives is in doubt as several of the more concrete proposals in the Commission draft were not included in the Presidency Conclusions (EEB, 2001) and as the SDS "still has a strong focus on economic policies, as reflected by the economic bias of the Spring Reports" (EEAC, 1993: 2). The European Environmental Advisory Council further notes that the SDS "...has an additive rather than integrative character". The strategy is scattered over several documents and lacks an accessible and understandable message, which considerably weakens its impact on everyday decision-making. This format does not stimulate public participation. Its messages have received extremely limited public attention, and even among small expert circles its influence has not been great. Accordingly, the EU-SDS does not provide the necessary leadership and its impact on public debate is limited. So far, the EU-SDS has not strengthened civil society engagement (EEAC, 2003: 2). Whether the SDS process can provide significant impetus to the integration agenda will depend on the effectiveness of the proposals related to improving governance for sustainability, especially the sustainability impact assessment.

Impact Assessment

In 2002, the European Commission launched the new Impact Assessment (IA) procedure. The change of name from "sustainability impact assessment" to "impact assessment" reflects the merging of two related but distinct policy initiatives into one procedure:

• The idea of sustainability impact assessment as referred to in the Gothenburg Presidency Conclusions was developed as an instrument to implement the EU SDS. The idea behind sustainability impact assessment is that sectoral policy initiatives tend to be designed on the basis of a narrow evaluation of their potential effectiveness in solving the given policy problem weighed up against direct costs. To broaden the scope of ex ante evaluation, the procedure is designed to give more consideration to potential secondary effects of a sectoral policy on environmental services and quality. In this sense, sustainability impact assessment presents a shift of emphasis from generic integration mechanisms to a procedure that takes specific policy initiatives as its starting point.

- At the same time, the Lisbon process and the Governance White Paper (CEC, 2001a) have led to a review of the EU regulatory impact assessment procedures. These include ex ante evaluations that are legally required under certain circumstances (budgetary evaluation, business impact assessment and environmental impact assessment) as well as those that are not (e.g. gender assessment, trade impact assessment, and SME assessment). The review aims to simplify regulation, avoid unnecessary rules, reduce regulatory burdens for business and increase transparency. The idea behind this review is that more care is to be taken to ensure that the benefit of regulation is proportionate to its regulatory burden and that regulation can be significantly streamlined.

This dual objective of IA is clearly expressed in the Commission Communication: "The Commission intends to launch impact assessment as a tool to improve the quality and coherence of the policy development process. It will contribute to an effective and efficient regulatory environment and further, to a more coherent implementation of the European Strategy for Sustainable Development" (CEC, 2001b). The procedure is being introduced gradually throughout 2003 and 2004 but will become mandatory for all major policy proposals – whether regulatory initiatives, financial interventions or "softer" instruments. It replaces previously separate regulatory impact assessments and consists of a preliminary assessment for all proposals and an extended assessment for those that are thought to have "substantial economic, environmental and/or social impacts" (CEC, 2001b).

IA aims to generate and gather knowledge to make a better-informed decision, as well as opening up the decision process to a range of stakeholders and increasing transparency. The assessment is carried out by the Directorate-General responsible for the policy proposal during the process of decision-making. The assessment needs to be documented in an Impact Assessment Report, which covers: problem identification; objective of the proposal; policy options; impacts; further analysis; and follow-up.

It is too early to judge the effectiveness of the EU Impact Assessment as an instrument for environmental policy integration because the procedure has just begun to be rolled out, and training, tools and guidance documents are only beginning to become available. However, a preliminary contents analysis of the first 20 Extended Integrated Assessment Reports reveals that many of these early assessments are of modest quality (Hertin et al., 2004). For example, many only consider one policy option, are narrowly focused on direct economic impacts and explore social and environmental impacts only briefly, some are not evidence-based, and few are transparent about how and by whom the assessment has been carried out.

Environmental Policy Integration in the EU

The last ten years have seen a range of attempts to address the impacts EU policies have on the environment. Ambitious environmental regulations are often helping improve the quality of the European environment, but agricultural, fishing, transport, and energy policies tended to create and support structures that are in direct conflict with environmental objectives. The launch of four major initiatives in quick succession – internal communication mechanisms, Cardiff process, Sustainable Development Strategy and Impact Assessment – may give the impression that EU policy-makers are now compensating for lack of real progress with action. The first part of the chapter, however, showed that environmental policy integration has a number of distinct dimensions which need to be addressed by specific procedural and organisational change. The EU policy integration initiatives are legitimately targeting different barriers to integration: lack of communication (internal integration mechanisms); missing top-down support (SDS); insufficient capacity and agenda setting (IA and Cardiff process). On the other hand, critics are right to point out that progress is hindered by the fact that these initiatives have been developed in what appears to be an unco-ordinated process characterised by hasty changes and superficial compromises. Integrating the SDS into the Lisbon process was a major step to improve coherence, but it is still unclear exactly how the Cardiff process, the Sixth Environmental Action Programme and the IA procedure fit into this bigger picture.

The greater problem is that environmental policy integration at the EU level is dependent on better integration in member states. This is not only because the complex multi-level structures of EU policy-making make the challenge of integration more difficult in the "technical" sense of co-ordinating positions and policy approaches. Member states hold much of the power in EU policy-making, particularly through the Council of Ministers which has a very fragmented structure and which some see as the real barrier to integration in the EU (Kraack et al., 2001). They are also responsible for implementation, which needs to be regarded as an extension of the policy-making process (Jordan, 1999). Unless national governments improve their internal coherence, the political capacity of the Commission to promote integration should not be over estimated.

Conclusions: Assessment and Prospects

The experience of the EU shows that environmental policy integration, while signalled in cornerstone EU legislation, such as the Single European Act and the Amsterdam Treaty, and promulgated by a welter of new functions and procedures, has proven difficult in practice. We have reviewed a number of explanations for the slow progress. First, it has been argued that it may be linked to the mismatch between traditional modes of organising administrations into specialised departments and the objective of cutting across the boundaries and interests this generates. This structural analysis would suggest that policy integration can only succeed with much broader reform of administrations, perhaps into more adaptive project-based forms, as has occurred in some industries. Second, others have argued that policy integration should be seen as a convenient rhetorical position taken up at EU intergovernmental level that has little substance. In this reading, policy integration is merely symbolic politics, designed to neutralise environmentalist critiques of EU policy, while permitting established sectoral domains to continue to operate more or less unchanged. The accretion of administrative function and procedure around policy integration serves, in this interpretation, merely to obfuscate business-as-usual.

Third, and paradoxically, policy integration has also been seen as a deliberate strategy to water-down environmental policy. Integration may be seen as a means to encourage the adoption of voluntary and "soft" modes of environmental governance by turning away from more legislative routes. In becoming devolved into the activities of sectoral departments, questions of environmental appraisal and protection may be compromised on more easily than if they had been subjected to higher-level and more intensive scrutiny (Lenschow, 2002). It may also be that, as we have seen in practice, the very procedures designed to enable policy integration create new incentives and opportunities to resist environmental considerations being taken into account in sectoral policies, by opening the door for other policies to challenge environmental initiatives (Kraack et al., 2001).

A fourth policy failure explanation is that the process of integration has not yet been pursued in a strategic and co-ordinated way; that it has yielded a range of well-meant, but ad hoc arrangements that are, as yet, inadequate to the task. In this analysis, insufficient attention has been paid to the structural and interest-based problems that will be faced by any process of environmental policy integration, so that measures aimed at achieving it have been poorly designed and positioned. Strategies and mechanisms have often been unsuccessful because other policy sectors have found ways to reduce those features they see as threats to their own interests and autonomy. Procedures – whether Impact Assessment, Cardiff requirements or Green Star requirements – may be complied with in some minimal way, but this does not produce policy integration.

Strategies that involve sectoral policy administrations directly (such as the Cardiff process and the Impact Assessment procedure) aim to overcome the problem of sectoral resistance by exchanging a top-down, procedurally-driven process with a bottom-up and consultative process which starts with the particularities of a policy domain. These strategies are based on a principle of deliberation as the basis for learning, rather than the exercise of bureaucratic power. It gives policy actors an active role in framing the integration agenda for themselves, while seeking to challenge the traditional assumption of a trade-off relationship between sectoral and environmental aims. Rather than trying to change behaviours through the mechanism of new procedures, they aim to integrate environmental concerns into the life-blood of different parts of government. The purpose of guidance and performance indicators is to seek to avoid disparities in outcomes and the threat of "dilution" that may result.

Although the translation of these insights into lessons for policy remains subjective, we would like to suggest a number of more general conclusions:

1. *Integration must be integrated*: A successful integration effort needs to involve a mix of strategies. It should combine bottom-up with top-down mechanisms and address all four functions of integration in a co-ordinated and mutually-reinforcing way (agenda setting, horizontal co-ordination, capacity building and policy learning).
2. *Integration involves learning*: It would be unrealistic to conceive an integration strategy as a list of short-term measures to be implemented according to a set timetable. Instead, it needs to be understood as an ongoing, long-term process consciously designed to promote internal capacity and policy learning, for example through appropriate internal evaluation and feedback mechanisms. Learning is also more likely when policy actors can be rewarded for positive achievements.
3. *Integration requires evaluation*: Monitoring and reporting are needed as a top-down instrument which allows external stakeholders (including the environment administration) to assess progress towards integration. Quantitative and qualitative indicators are needed because integration is an ambiguous concept and success claims are difficult to validate without them.
4. *Integration thrives with leadership*: The active involvement of the central government is key to success. Without a certain level of commitment from the executive core, integration mechanisms may lead to a weakening rather than a strengthening of environmental policy. Leadership needs to be exercised through declaratory postures and by good example.
5. *Integration requires realism*: Sector-specific approaches need to clarify the key tension-points in processes of environmental policy integration. These may be structural, and linked to the standing and autonomy of the administration as a whole, or they may relate to policy options for which trade-offs do need to be made in the absence of a satisfactory convergence between social, economic and environmental objectives. Without a dispassionate approach to the limits of integration, the whole project risks being little more than gestures and missed opportunities. To ensure coherence between sectoral strategies, they should be related to an overarching environmental strategy endorsed by the whole of government.

We believe that it is still too early to judge which of the contending explanations of current moves towards greater environmental policy integration in the EU is correct. On balance, the number of initiatives that have been launched in the recent past, and the developing logic of high-level commitments to sustainable development, would argue for a more positive reading of the phenomenon of integrationism. While these initiatives are yet to bear much fruit, they are incrementally creating an ever-greater number of opportunities for policy actors to press home environmental messages and considerations. The expectation must be that, over the longer term, some of these opportunities will be exploited.

156 New Modes of Governance

References

Berkhout, F. and Gouldson, A. (2003), 'Inducing, shaping and modulating: perspectives on technology and environmental policy', in Berkhout, F., Leach, M. and Scoones, I. (eds), *Negotiating Environmental Change*, Cheltenham: Edward Elgar, pp.233-262.

Blazejczak, J., Edler, D., Hemmelskamp, J. and Jänicke, M. (1999), 'Umweltpolitik und Innovation: Politikmuster und Innovationswirkungen im internationalen Vergleich', *Zeitschrift für Umweltpolitik und Umweltrecht*, **22**(1): 1-32.

CEC (Commission of the European Communities) (1998), *Partnership for Integration – A strategy for integrating environment into EU policies*. COM(98) 333. Brussels: CEC.

CEC (2001a), *European Governance: A White Paper*, COM(2001) 428 final, Brussels, 25 June.

CEC (2001b), *A Sustainable Europe for a Better World: A European Strategy for Sustainable Development*, COM(2001) 264 final, Brussels, 15 May.

EEAC (2003), *Strengthening sustainable development in the EU*, recommendations from the EEAC Working Group on Sustainable Development as a contribution to the review of the European Sustainable Development Strategy and the preparation of the EU Spring Council in 2004. European Environmental Advisory Councils.

EEB (European Environmental Bureau) (2001), *Environmental Policy Integration – Proposals for a better institutional framework, based on an examination of the Cardiff Process*, Brussels: EEB.

Fergusson, M., Coffey, C., Wilkinson, D., Baldock, D., Farmer, A., Kraemer, R. A. and Mazurek, A.-G. (2001), *The effectiveness of EU council integration strategies and options for carrying forward the 'Cardiff' process*, Institute for European Environmental Policy.

Freeman, C. (1996), 'The greening of technology and models of innovation', *Technological Forecasting and Social Change*, **53** (1, September 1996): 27-39.

Gouldson, A. and Murphy, J. (1996), 'Ecological modernization and the European Union', *Geoforum*, **27**(1): 11-21.

Hertin, J. and Berkhout, F. (2003), 'Analysing institutional strategies for environmental policy integration: The case of EU enterprise policy', *Journal of Environmental Policy and Planning*, **5**(1): 39-56.

Hertin, J., Bartolomeo, M., Giugni, P., Jacob, K., Volkery, A., Wilkinson, D. and Zanoni, D. (2004), Review of 2003 Extended Impact Assessments – Preliminary conclusions. Unpublished draft.

Jänicke, M. (1990), *State failure: the impotence of politics in industrial society*, Cambridge: Polity Press.

Jordan, A. (1999), 'The implementation of EU environmental policy: a policy problem without a political solution?', *Environment and Planning C: Government and Policy*, **17**(1): 69-90.

Jordan, A. and Lenschow, A. (2000), '"Greening" the European Union: What can be learned from the "Leaders" of EU environmental policy', *European Environment*, 10(3): 109-120.

Jörgens, H. (1996), Die Institutionalisierung von Umweltpolitik im internationalen Vergleich. *Umweltpolitik der Industrielaender*, M. Jänicke (ed.), Berlin: Edition Sigma: 59-112.

Kemp, R. and Rotmans, J. (2001), *The management of the co-evolution of technical, environmental and social systems*, Towards Environmental Innovation Systems Conference, Garmisch-Partenkirchen.

Kraack, M., Pehle, H. and Zimmermann-Steinhart, P. (2001), Umweltintegration in der Europaeischen Union – Das umweltpolitische Profil der EU im Politikfeldvergleich. Baden-Baden: Nomos Verlagsgesellschaft.

Kraemer, R. A. (2001), Ergebnisse des "Cardiff-Prozesses" zur Integration der Erfordernisse des Umweltschutzes in andere Politiken – Bewertung des Zwischenstandes. Ecologic – Berlin: Institute for International and European Environmental Policy.

Lebessis, N. and Paterson, J. (1998), A learning organisation for a learning society: Proposals for "Designing Tomorrow's Commission". European Commission, Brussels: Forward Studies Unit.

Lenschow, A. (ed.) (2002), *Environmental Policy Integration*, London: Earthscan.

Nollkaemper, A. (2002), Three Conceptions of the Integration Principle in International Environmental Law, *Environmental Policy Integration*, A. Lenschow (ed.) London: Earthscan: 22-32.

Pressman, J.L. and Wildavsky, A. (1979), *Implementation: how great expectations in Washington are dashed in Oakland*, Berkeley, Calif ifornia and London: University of California Press.

Sabatier, P.A. and Jenkins-Smith, H. (1988), *Policy change and policy-oriented learning: exploring an advocacy coalition framework*, Dordrecht and London: Kluwer.

Schepelmann, P. (2000), *Von Helsinki nach Goeteborg*, Vienna: Sustainable Europe Research Institute.

Schout, A. and Jordan, A.J. (2005), Coordinated European Governance: Self-organizing or Centrally Steered? *Public Administration* (in press).

Wallace, D. (1995), *Environmental policy and industrial innovation: Strategies in Europe, the USA and Japan*. London: Royal Institute of International Affairs Energy and Environmental Programme/Earthscan.

Weale, A. (1992), *The new politics of pollution*, Manchester and New York: Manchester University Press.

Wilkinson, D. (1997), 'Towards sustainability in the European Union? Steps within the European Commission towards integrating the environment into other European Union policy sectors', *Environmental Politics*, **6**(1), 153-173.

Wilson, J.Q. (1980), *The Politics of Regulation*, New York: Basic Books.

Chapter 9

The Challenges of Policy Integration from an International Perspective: The Case of GMOs

Joseph Murphy and Joanna Chataway

In the early morning of 25 June 1999, the Environment Council of the European Union (EU) imposed a de facto moratorium on the authorisation of new genetically modified crops in the EU. This action followed widespread public protests against genetically modified organisms (GMOs) in Europe. The German President at the time, Juergen Trittin, explained the action as follows: "There was a political consensus that everything should be done to avoid any authorisations under the existing rules." However, acknowledging the fact that the legal status of the action was ambiguous, Trittin continued: "It will be a de facto moratorium, though legally speaking we can't call it that" (FoEE, 1999).[1]

A trade conflict with the United States (US) is probably the most significant problem the EU faces as a result of the moratorium. In the years that followed the Environment Council's decision, the US continued to authorise and plant new GMOs. However, because of the bulk handling of agricultural commodities in the US, it became increasingly likely that exports to the EU would contain "illegal" GMOs – ones not authorised in the EU. To cope with this problem, commodity traders in Europe began to source products from outside the US. South American producers, in particular, benefited from the shift. US officials have blamed a decrease in exports on the moratorium. In May 2003 the US made a formal complaint to the World Trade Organization regarding the EU's failure to approve any new GM crops over the previous five years (for an overview of the conflict see Pew, 2003).

This chapter uses the EU-US conflict over GMOs to analyse how values and interests shape the use of science in international agreements and institutions. From this perspective it draws out implications for governance and policy integration at the international level. The chapter focuses on the international agreements and institutions that deal with the possible environmental or human health risks associated with trade in GMOs. These are:

---

[1] At the time of writing, the status of the moratorium appears to be in a state of flux.

- the Sanitary and Phytosanitary Measures Agreement of the World Trade Organization;
- the global food standards of the FAO/WHO Codex Alimentarius Commission;
- the Biosafety Protocol to the UN's Convention on Biological Diversity.

This chapter will show how, with the emergence of the EU-US conflict over GMOs, these agreements and associated institutions were drawn into a complex international governance process. This case raises doubts about some of the assumptions made in the governance literature, for example the existence of "a problem" or "a goal", the definition of which is widely accepted, and around which collective action can take place. It also raises doubts about linking governance with the goal of policy (or institutional) integration. This case suggests that in some cases lack of integration is the outcome of governance. And, in the case of GMOs at the international level, lack of integration of agreements and institutions played an important role in the "management" of a conflict for a period of time, because of the political possibilities created by it. This chapter begins with some observations on globalisation, international agreements and the role of science.

Globalisation and International Agreements: What Role for Science?

Globalisation is one of the most widely debated phenomena of the late 20th and early 21st centuries. Definitions of the concept and its use are contested. However, if it is taken to mean the tendency of the world to become more integrated over time with increasing speed, it is undeniable. And it is a useful starting point for this chapter because international agreements are one of the main ways in which governments attempt to manage integration at the global level and in doing so impact on globalisation processes more generally. Put simply, international agreements establish the rules (and related institutions) that govern how different jurisdictions and cultures should interact with each other.

The market – and particularly how it should be controlled – is one of the most important integration and globalisation issues. There are significant differences of opinion over the way in which market integration should proceed. Many of the arguments are not new. For example, Polanyi, writing in the 1940s, characterised the political economy of the early 20th century as revolving around efforts to establish the "utopia" of a self-regulating market economy (Harriss, 2000; Polanyi, 1944). Polanyi felt that this was an historical watershed. "While history and ethnography know of various kinds of economies, most of them comprising markets, they know of no economy prior to our own, even approximately controlled and regulated by markets" (Polanyi, 1944, p.44).

For a short while, during the late 1980s and early 1990s, as the USSR collapsed and neo-liberal market economics gained ground, it seemed that at least the principles – if not the practice – of the self-regulating market economy might be accepted as the norm. Francis Fukuyama, a former US government analyst, dramatically proclaimed the "end of history" (Fukuyama, 1992). As Charles Gore noted, however, "The end of history lasted for such a short time". The East Asian crisis, the rise of anti-globalisation protests in Seattle and the events of 11 September 2001 all in different ways seemed to signify a new era of debate about the role that markets should play in the social, economic and political order (Gore, 2000, p.1084).

Today we can begin to understand the role of international agreements in managing processes of integration and globalisation by recognising that most (if not all) agreements have an economic and social objective. The economic objective of international agreements is to improve the efficiency of the market system. Advocates of market efficiency justify this by arguing that related increases in economic activity will result in increases in welfare. International agreements in the areas of trade liberalisation and intellectual property rights are examples. The social objective of international agreements, however, is to ensure that the reality of international integration is consistent with the needs, preferences and expectations of those involved. Examples include multilateral environmental agreements (MEAs) and agreements on human rights. These economic and social objectives have been described as the different "rationalities" that underpin international agreements (Isaac, 2002).

What role does (should) science play in international agreements? In practice, science plays a number of different roles. In many cases, for example, it provides the evidence base on which an international agreement is built. Many MEAs can be given as examples. In most situations, without a scientific consensus on the problem, it is unlikely that an international agreement will be concluded. However, evidence and consensus is not a necessary or sufficient basis on which to build an international agreement. The "precautionary principle", for example, is a way of justifying action without clear scientific proof that there is a problem. An example is the Montreal Protocol, which was negotiated successfully in the absence of a clear scientific understanding of the causes of ozone depletion. Conversely, the withdrawal of the United States from the climate change convention and the Kyoto Protocol demonstrates that scientific consensus will not necessarily result in international action.

More generally science has become a standard on which many international agreements are based. It is deployed enthusiastically at the international level because it is seen as a way of bridging between different jurisdictions and cultures. Science offers the possibility of revealing (proving) and overcoming divergent practices at the national level. It is, some would say, a common language on which we can build a level playing field. One of the areas where this is particularly important is traded goods and risk. Advocates of free trade only accept a limited number of reasons why a government might legitimately restrict trade. One of these is scientific proof that a product poses a risk to human health or the environment.

To what extent is science able to play these roles in international agreements? Scholars who study science from a social science perspective have identified a number of problems. For example, how can uncertainty and ignorance about risk be dealt with in an objective and scientific manner? In the past, it might have been possible to depend on scientific consensus and peer review processes but increasingly the independence of scientists is being called into question. And, even if this was not the case, so-called "non-scientific" issues still influence the most independent of scientists when they make decisions about uncertainty and risk. As Jasanoff (1993, p.129) has argued:

> We can hardly order, rearrange, or usefully supplement our knowledge about risk without incorporating these issues into a clear, framing vision of the social and natural order that we wish to live in.

Problems such as these alert us to the various ways in which values and interests might impact on the use of science in international agreements. In the following three sections, as an example, we look more closely at the regulation of genetically modified organisms (GMOs) at the international level. The evidence clearly shows that the SPS Agreement of the World Trade Organization, the food standards of the Codex Alimentarius and the UN's Biosafety Protocol treat the problem of uncertainty in risk assessment differently. This can be explained to some extent by taking into account the values and interests involved and it has important implications for the governance and the goal of policy integration.

The World Trade Organization and GMOs

The multilateral trading system began to emerge in 1948 with the General Agreement on Tariffs and Trade (GATT). Successive trade talks built on this agreement and led to the launch of the Uruguay round of trade negotiations in 1986. These negotiations concluded in 1993 and the World Trade Organization (WTO) was launched in 1995. The main purpose of the WTO is the implementation of trade law and the promotion of further trade liberalisation. It also provides member countries with a legal mechanism for dealing with breaches of trade rules. In this section we focus on the way the WTO deals with risk and scientific uncertainty. The WTO does not have specific provisions relating to genetically modified organisms but the Agreement on Sanitary and Phytosanitary Measures and the dispute resolution process are important.

The Agreement on Sanitary and Phytosanitary Measures

It is often claimed that trade liberalisation puts human health and the environment at risk because trade law makes it difficult to exclude risky products from markets. However, trade law actually defines the circumstances under which trade rules can be set aside to protect these. With respect to trade in GMOs, the Agreement on the Application of Sanitary and Phytosanitary Measures (SPS Agreement) is particularly important. This agreement was concluded during the Uruguay round of negotiations and it covers all measures that aim to protect human or animal health from food-borne or plant-carried risks. The SPS Agreement negotiations were not conducted with the problem of trade in GMOs in mind but from the mid-1990s the agreement became increasingly central to the GMO debate.

Essentially the SPS Agreement defines the circumstances under which trade in agricultural products can be restricted if they pose a threat to human or animal health. Science, as a basis for objective decision-making, is emphasised as follows (Article 2 Paragraph 2):

> Members shall ensure that any sanitary or phytosanitary measure is applied only to the extent necessary to protect human, animal or plant life or health, is based on scientific principles and is not maintained without sufficient scientific evidence...

Therefore, the SPS Agreement invokes science as an objective and apolitical basis on which to make decisions about the risks associated with trade in agricultural goods. It also establishes where the burden of proof falls – there must be evidence of risk. Significantly, the SPS Agreement also identifies the Codex Alimentarius Commission as a source of objective science-based standards in the area of food safety (see below).

How does the SPS Agreement deal with the problems of uncertainty and ignorance where risk is concerned? At least to some extent these issues are dealt with in Article 5 Paragraph 7:

> In cases where relevant scientific evidence is insufficient, a Member may provisionally adopt sanitary or phytosanitary measures on the basis of available pertinent information, including that from the relevant international organizations as well as from sanitary or phytosanitary measures applied by other Members. In such circumstances, Members shall seek to obtain the additional information necessary for a more objective assessment of risk and review the sanitary or phytosanitary measure accordingly within a reasonable period of time.

It is important to note that, although the problem of uncertainty is acknowledged, the words "provisionally" and "reasonable period of time" imply that uncertainty about risk will be a short-lived problem. The article implies that further research will clarify any outstanding questions.

On 13 May 2003 the US made a formal complaint to the WTO regarding the EU's moratorium on GMOs. In her letter, Linnet Deily, the US Ambassador to the WTO, identified two key problems: "[1] Since October 1998, the EC has applied a moratorium on the approval of biotech products... [2] Moreover, the Member States maintain a number of national marketing and import bans on biotech products even though those products have already been approved by the EC..." The Ambassador went on to state that this placed the EU in violation of various pieces of trade law. Articles 2 and 5 of the SPS Agreement, as discussed above, were given first in the list. Although there is no way of knowing whether or not the WTO dispute process will be allowed to reach a conclusion, and if it is what the decision will be, this action placed the SPS Agreement (and its meaning) at the centre of the current conflict.

The WTO and Dispute Settlement

The second area of trade law that is important regarding the EU-US conflict over GMOs and risk is case law. This accumulates as cases go through the WTO's dispute settlement process and the WTO is forced to interpret and apply trade law. Past cases indicate how trade law might be interpreted in the future in certain situations. Whether they were won or lost influences the subsequent thinking of trade officials and others. Following a number of high profile cases there is a better understanding of how trade law is interpreted in situations involving the protection of human health and the environment when there is scientific uncertainty. Perhaps the most important case, which impacts on the EU-US conflict over GMOs in various ways, is the so-called beef-hormone dispute.

The beef-hormone dispute began with Europe's ban on imports of hormone treated beef from North America in 1981. The ban was defended on the grounds that the hormones posed a threat to human health. The regime was extended further in 1988 and 1996. For much of this time the US, who argued that the action was illegal, imposed retaliatory trade sanctions. Significantly, the dispute ran for such a long time partly because the parties could not agree on how the scientific issues should be dealt with under the GATT system that existed before the Uruguay round of trade negotiations was completed. However, during the Uruguay round, the US government pursued the negotiation of the SPS Agreement (see above) to clarify how the WTO should deal with this kind of science-based trade dispute. More specifically, the SPS Agreement was negotiated as a basis from which to challenge the EU's beef-hormone regime.

When the case was eventually judged the Appellate Body of the WTO found that the EU's measures were incompatible with the SPS Agreement. This conclusion was reached despite the fact that the beef-hormone regime was not found to discriminate against US products or to represent a disguised barrier to trade. In coming to its conclusion the Appellate Body focused on Article 2 Paragraph 2 of the SPS Agreement (see above). They pointed out that this allows countries to implement SPS measures but only if these are based on scientific principles and not maintained without sufficient scientific evidence. The main weakness of the EU's position, they ruled, was that the beef-hormone regime was not underpinned by a focused risk assessment. The Appellate Body decided that it was informed only by theoretical or hypothetical risks.

This is an important judgement in trade law but it is not easy to draw clear conclusions from it. From one perspective it might appear to be a clear endorsement of the so-called "sound science" perspective. As stated by Scott (2000, pp.148-149):

> It is crucial to an understanding of this Agreement to appreciate the emphasis which it places upon science and scientific reason. Measures ... in order to be compatible with the Agreement, must be susceptible to justification in the language of science.

However, it is also true that the Appellate Body's judgement focused on the process that the EU had gone through before establishing the beef-hormone regime and not the science involved. They drew attention to a procedural problem – the lack of a focused risk assessment. Other similar judgements have also focused on the failure of countries to follow a process before implementing trade restrictive measures. This is relatively easy to explain. For obvious reasons it is easier for the WTO to judge a case on process grounds rather than making its own assessment of risks where experts might be giving different and often contradictory advice. The WTO's dispute settlement process is not designed to settle disputes about science. It is designed to settle disputes about trade.

To conclude this section it is worth noting the extent to which some of these agreements, institutions and interactions are related to the interests of those involved. It has been argued convincingly by various critical commentators that trade liberalisation in its current form operates in the interests of some and not others (e.g. Stiglitz, 2002). One example of this is agriculture. Both the US and the EU protect their agricultural sectors in ways that are inconsistent with the basic tenets of trade liberalisation. They are also major exporters of related raw and processed products. With this in mind it is easier to understand their enthusiasm for negotiating the SPS Agreement during the Uruguay round. It established rules that made it very difficult for countries to restrict agri-food imports. On GMOs more specifically the US challenge to the EU at the WTO can be linked to a loss of as much as $300 million dollars per year of corn exports to the EU (Pew, 2003). However, more generally, the challenge was motivated by the fear that other countries might adopt the EU's position on GMOs. This would create a serious problem for biotechnology and agriculture in the US. Debates about the role that science plays in international agreements and about harmonisation and integration have to be seen in light of these broader political influences.

The Codex Alimentarius Commission and GMOs

The Food and Agriculture Organization (FAO) and the World Health Organisation (WHO) created the Codex Alimentarius Commission (CAC or Codex) in 1961. Its role is to develop global food standards – quality and safety – and it currently has over 150 members (countries). Committees prepare draft standards and traditionally these only become part of the Codex Alimentarius – the food code – after an eight stage process if there is a consensus amongst the members (Kimbrell, 2000). Various aspects of the work of Codex have implications for the regulation of GMOs at the international level. In this section we focus on Codex's attempts to develop generic principles and specific rules for risk assessment of GM foods.

Generic Principles of Risk Assessment

At its 22nd meeting in 1997 the Codex Alimentarius Commission adopted an 'Action Plan for Codex-wide Development and Application of Risk Analysis

Principles and Guidelines' (CAC, 1997). This document instructed the Codex Committee on General Principles (CCGP) to "...elaborate integrated principles for risk management and risk assessment... for inclusion in the Procedural Manual". In 1999, the CAC confirmed its instruction to the CCGP when it adopted a Medium Term Plan (1998-2002) which stated:

> ...the integration of risk analysis principles into the Codex decision-making process should be completed in the period by introducing necessary changes to... the General Principles of the Codex Alimentarius [amongst other things]... Specific guidance on the application of risk analysis principles should be provided to Codex Committees on one hand and to Member Governments on the other: the former guidance to be included in the Procedural Manual, the latter in the Codex Alimentarius itself. (CAC, 1999)

This statement makes it clear that the intention, at least at this stage, was that the output from the CCGP's work would be included in the Codex Alimentarius as a global standard.

In practice, the CCGP found it extremely difficult to agree on risk analysis principles. At the 15th session of the CCGP (10-14 April 2000) they encountered problems at two levels, although both involved the problem of how to deal with uncertainty in risk assessment (CAC, 2001a). At the more abstract and rhetorical level the CCGP became embroiled in a debate about the "precautionary principle" (PP). They found it impossible to achieve a consensus. The European Union and others argued for the inclusion of the PP but the United States refused to accept that such a principle exists. The US argued that the PP involves politics and not science (see Dratwa, 2002). More specifically, in the area of risk management, there were different opinions on the working text. The draft from this meeting included the following bracketed alternative proposals for paragraph 34:

> 34. [Where relevant scientific evidence is insufficient, precaution can be exercised as an interim measure to protect the health of consumers. However, additional information for a more objective risk assessment should be sought and the measure taken reviewed accordingly within a reasonable timeframe.]

> 34. [When relevant scientific evidence is insufficient to objectively and fully assess risk from a hazard in food, and where there is reasonable evidence to suggest that adverse effects on human health may occur, but it is difficult to evaluate their nature and extent, it may be appropriate for risk managers to apply precaution through interim measures to protect the health of consumers without awaiting additional scientific data and a full risk assessment, in accordance with the following criteria:]

These alternatives are interesting for various reasons. The first (a proposal from Malaysia) reveals the extent to which the wording of the WTO's SPS Agreement was impacting on the discussions (compare with Article 5 Paragraph 7 of the SPS Agreement given above). The second (a joint US and EU proposal) shows the accommodation which had been reached at this stage regarding the precautionary principle. A footnote to the second alternative paragraph 34 states: "Some members refer to this concept as the 'precautionary principle'."

A special working group was set up to deal with the paragraph 34 problem. The working group developed a text and submitted it to the 16th Session of the CCGP. Some members were unhappy with it. The US, for example, expressed general reservations on the whole text. The 16th Session of the CCGP then became embroiled in what proved to be a fruitless attempt to resolve differences of opinion on uncertainty and the precautionary principle. Some delegations argued that, regardless of its scientific merit, the precautionary principle is important for public confidence. Realising that the WTO/SPS link to Codex was problematic, other delegations wondered if it would be possible to negotiate something that was less significant than a Codex standard. Ultimately the 16th Session of the CCGP ended with more proposals for paragraph 34 rather than fewer (CAC, 2001b). Agreement on a complete text was impossible. The CCGP agreed to seek further advice from the Codex Alimentarius Commission, particularly on "the action to be taken when scientific data were insufficient" (CAC, 2001c). In doing so the CCGP accepted that it would not reach a conclusion by the deadline that had been set for it.

At the 24th Session of the Codex Alimentarius Commission (2-7 July 2001) the initial mandate of the CCGP was reconfirmed. At the same time many of the arguments which had becalmed the CCGP were rehearsed again, albeit this time in a more significant setting – a meeting of the CAC itself. In an attempt to reach a compromise, the Chairperson proposed that the CAC should take the following position on scientific uncertainty:

> When there is evidence that a risk to human health exists but scientific data are insufficient or incomplete, the Commission should not proceed to elaborate a standard but should consider elaborating a related text, such as a code of practice, provided that such a text would be supported by the available scientific evidence. (CAC, 2001c, 12 (para. 81))

With this proposal the CAC accepted that the problem was too difficult and that negotiating a Codex standard made the stakes too high. However, 17 countries, all of them European apart from Sudan, objected to this proposal. They argued that this move would ultimately prevent Codex from addressing risks to human health. Despite these objections the Codex Alimentarius Commission adopted this proposal by a majority vote. This prompted the UK – not one of the objectors to the proposal – to complain on the grounds that "it was essential to make decisions by consensus at the level of the Commission" (CAC, 2001c, 12 (para. 84)).

Although it was unsatisfactory in many ways, the decision taken at the 24th Session of the Codex Alimentarius Commission did result in the CCGP making progress in 2002-03. Arguably this was achieved by simply avoiding the problematic issues. The "related text" option meant that any outcome would not, as was intended originally, be included in the Codex Alimentarius and therefore it would not have that status in any future trade dispute at the WTO. However, despite all these problems, the "Working Principles for Risk Analysis for Application in the Framework of the Codex Alimentarius" were adopted at the 26th Session of the Codex Alimentarius Commission (30 June – 7 July 2003). The compromise text given above was included as paragraph 10. Interestingly, the delegation from Italy, speaking on behalf of the European Union, then expressed the view that paragraph 10 – on what to do when there is insufficient scientific data – means that there should now be a review of all current Codex food safety standards. Others disagreed but the Commission stated that "existing standards may be reassessed on a case-by-case basis" (CAC, 2003a, 20 (para. 144)).

Risk Assessment of GM Foods

The lack of Codex standards specific to GM foods was raised at the 23rd Session of the Codex Alimentarius Commission in the summer of 1999. At this meeting the US proposed a "Biotechnology Code" which would have reiterated existing Codex statements on science-based standards. This proposal was rejected and as an alternative the Codex Alimentarius Commission agreed the terms and conditions for an Ad Hoc Intergovernmental Task Force on Foods Derived from Biotechnology (the Task Force). The objective of the Task Force was:

> To develop standards, guidelines or recommendations, as appropriate, for foods derived from biotechnology or traits introduced into foods by biotechnology, on the basis of scientific evidence, risk analysis and having regard, where appropriate, to other legitimate factors relevant to the health of consumers and the promotion of fair trade practices. (CAC, 2003a, p.108)

The Task Force was instructed to submit a full report in 2003.

Various aspects of the Task Force's work are relevant to this discussion but its work on allergenicity is particularly useful. By focusing on this we can examine how uncertainty in a specific area has been dealt with. Put simply, the problem of allergenicity in food safety assessment is challenging because no single test will definitively establish whether or not a food will cause an allergic reaction in a consumer. There are, instead, a variety of tests that can be used. Each one of these will give some useful information. This is a problem for the safety assessment of GM foods because the process of genetically modifying an organism might result in the production/inclusion of a gene that codes for a protein that will result in a food allergy reaction. However, at the same time, there is no way of knowing for certain. The question then arises: How should the results of a number of tests, none of which is definitive, be dealt with?

By the time the Task Force met for the second time in March 2001 a well developed text already existed at Step 5 of the Codex procedure on safety assessment of foods derived from recombinant-DNA plants. The assessment of allergenicity was covered. The text recommended that:

A decision-tree strategy should be applied in the assessment of the potential allergenicity of the newly expressed protein(s). (CAC, 2001d, p.49 (para. 40))

In effect this would mean that a single test indicating a problem could be enough for a new GM food to fail a risk assessment. In a footnote the document defends this approach and states: "Decision tree strategies have been developed and modified on the basis of expert consultations in national and international fora..." (CAC, 2001d, footnote 5). Joint FAO/WHO expert consultations in 2000 and 2001 are given as examples.

Despite this apparent consensus, however, the approach of the Task Force to allergenicity was about to change. A separate annex on allergenicity was tabled at the 3rd meeting of the Task Force in 2002 (4-8 March), only a year later, and this took a different approach to the problem (CAC, 2003b). Put simply the decision tree approach had been replaced by what was now being called "the preponderance of evidence" approach. The change is explained: "This approach takes into account the evidence derived from several types of information and data since no single criterion is sufficiently predictive." The Canadian Chair of the Working Group defended the decision:

He noted that the Joint FAO/WHO Expert Consultation in January 2001 provided a valuable source of expert input for the Working Group to draw upon... The Working Group had discussed the outcome of the FAO/WHO Expert Consultation but come to the conclusion that it was not possible scientifically to arrive at clear "Yes/No" decisions at each and every step in the decision process. It had therefore recommended a more holistic approach that took into account a broad range of information... (CAC, 2003b, 11 (para. 62)).

This change of approach was controversial. Critics pointed out that in March 2001 the Task Force had a working document that recognised the FAO/WHO expert consultation of January that year as establishing a consensus on the subject. This underpinned the adoption of the decision tree approach. They asked why, only a year later, the allergenicity working group had come to a different conclusion which led them to choose the preponderance of evidence approach.

Perhaps not surprisingly any effort to explain this decision must go beyond the science to consider the values and interests of those involved. In practice, these two approaches to allergenicity might result in the same regulatory decision being made about the safety of a GM food. However, this will not necessarily be the case. Arguably the preponderance of evidence approach makes it easier for a GM food to pass a risk assessment because no single test that raises a doubt about allergenicity will result in it being rejected. This helps to explain why pro-GM stakeholders have supported this approach. Conversely, because the decision tree approach is more demanding, it is easy to understand why those stakeholders who take a more cautious or sceptical approach to the technology have supported this approach.

The Cartagena Protocol on Biosafety and GMOs

The United Nations Convention on Biological Diversity (UNCBD) was one of the agreements to emerge from the 1992 United Nations Conference on Environment and Development. The Convention aims to ensure "...the conservation of biological diversity, the sustainable use of its components and the fair and equitable sharing of the benefits arising out of the utilization of genetic resources...". Building on the Convention in 1995 the Conference of the Parties (COP) agreed to negotiate a Protocol on biosafety to deal with trade in "living modified organisms" (LMOs) and possible impacts on biological diversity. The Cartagena Protocol on Biosafety was finalised in January 2000 (in Montreal). It entered into force after the 50th ratification. In this section we discuss the Protocol, how it deals with risk issues and how it relates to other areas of international law. Differences of opinion that surfaced during the negotiations are examined.

The Cartagena Protocol on Biosafety establishes a framework for regulating trade in LMOs. The most important element is the "advance informed agreement" (AIA) procedure. AIA requires exporters of LMOs that will be released into the environment to provide a description of them to the importing country in advance of the first shipment. The importer must acknowledge receipt of this information within 90 days and then authorise the shipment within 270 days or state the reasons for rejecting it. The purpose of AIA is to allow a risk assessment to be reviewed or carried out to the satisfaction of the importer. Risk assessment is central to the working of the Protocol and it is covered in Article 15 and Annex III.

During the negotiations all parties agreed that risk assessment should be central to the Protocol. Consequently, disagreements on Article 15 and Annex III were less significant than on other issues. Article 15 of the Protocol states that:

> Risk Assessment undertaken pursuant to this Protocol shall be carried out in a scientifically sound manner... in order to identify and evaluate the possible adverse effects of living modified organisms on the conservation and sustainable use of biological diversity, taking into account risks to human health.

In the negotiations the most contentious aspect of this article was the reference to "risks to human health". Some negotiators wanted this to be removed. They argued that the Codex Alimentarius Commission was dealing with human health issues. However, widespread concerns about the risks associated with the technology, particularly in Europe, made it difficult to restrict the scope of the risk assessment. Also, the mandate from the 1995 COP made it clear that human health issues should be included (Andrén and Parish, 2002).

Annex III sets out the risk assessment methodology and the principles that should be taken into consideration. The principles are particularly important for interpreting the Protocol. The Annex states:

Risk assessment should be carried out in a scientifically sound and transparent manner... Lack of scientific knowledge or scientific consensus should not necessarily be interpreted as indicating a particular level of risk, absence of risk, or an acceptable risk.

This is one of the ways that the negotiators managed to deal with the problem of uncertainty. Andrén and Parish (2002, p.334) have argued: "This is a statement of fact, advising against jumping to conclusion, but otherwise it is of little practical help." However, it is significant that this statement represents a more balanced approach to the burden of proof compared to that which is included in the WTO's SPS Agreement (see above).

The relative ease with which risk assessment was dealt with can be contrasted with the controversy surrounding the inclusion of the precautionary principle in the Protocol. The final version of the Protocol contains two references to precaution as Rio Principle 15 (in the Preamble and in Article 1 (Objective)). To some extent these suggest a victory for those who argued in favour of the precautionary principle, particularly the European Union. However, during the negotiations explicit references to precaution were removed from other articles. Initially, the EU and others argued that it should be included in articles that deal specifically with decision-making because a framing reference is insufficient, i.e. one that appears in the opening articles. However, a group led by the US argued that this was unnecessary. The compromise allows countries to invoke precaution in a qualified way:

Lack of scientific certainty due to insufficient relevant scientific information and knowledge regarding the extent of the potential adverse effects of a living modified organism on the conservation and sustainable use of biological diversity in the Party of import, taking also into account risks to human health, shall not prevent that Party from taking a decision, as appropriate, with regard to the import of the living modified organism in question... in order to avoid or minimize such potential adverse effects. (Article 10 para. 6)

An almost identical paragraph is included in Article 11 on the procedure for LMOs intended for food, feed or processing (Bail et al., 2002, p.336).

Events that were unfolding well beyond the Biosafety Protocol negotiations help to explain how the PP is dealt with in the text. One of the participants commented as follows:

The readiness in Cartagena of all the negotiating groups to give up their positions on the precautionary principle for the sake of getting a final agreement could have weakened considerably the chance of getting the principle firmly anchored in the protocol at the next negotiating session. Surprisingly, this did not happen in Montreal. On the contrary, the inclusion of specific language on the precautionary principle became a more important issue, in part because of the continuous insistence by the Miami Group on the "savings clause" [see below] and because of the attempt to start a parallel process in the WTO on a market access regime for biotechnology products... Governments therefore became increasingly aware of the fact that precautionary measures in the area of LMOs could be subject to challenge through the WTO dispute settlement mechanism. (Graff, 2002, p.414)

This account draws attention to the impact of events that took place between the collapse of the Biosafety Protocol negotiations in Cartagena and the successful agreement of the Protocol in Montreal a year later. The way in which the Biosafety Protocol deals with scientific uncertainty can only be understood by taking into account the politics of trade and biotechnology at this time.

The "savings clause" mentioned in the above involved clarifying the relationship between trade rules and the Biosafety Protocol. Arguably, this is something that negotiators failed to do. The EU preferred not to address this problem at all because principles of international law emphasise more recent and more specific agreements over older and more general ones. However, the US argued in favour. The result is an uncomfortable compromise at the end of the Biosafety Protocol's Preamble. It states:

> Recognizing that trade and environment agreements should be mutually supportive with a view to achieving sustainable development,
> Emphasizing that this Protocol shall not be interpreted as implying a change in the rights and obligations of a Party under any existing international agreements,
> Understanding that the above recital is not intended to subordinate this Protocol to other international agreements...

This compromise wording is explained largely by the desire of the US and others to make sure that WTO rules are respected, whereas the EU and developing countries were keen to establish the Protocol as not subordinate to WTO rules (Afonso, 2002).

International Agreements, Science and Risk: Understanding the Role of Values and Interests

This chapter began with an outline of the EU's de facto moratorium on GMOs and the associated trade conflict with the US. We argued that this creates an opportunity to study international agreements and their use of science, and particularly how values and interests impact on this. It is also an opportunity to critically reflect on the ideas of "governance" and "integration".

In this chapter we have focused on risk assessment. This issue is particularly interesting because, at least at a rhetorical level, risk assessment is often presented as a purely scientific problem. Therefore, if we can expose values and interests impacting on the treatment of risk, it is likely that particularly interesting and important issues are being uncovered. Science is also often identified as an objective way of dealing with governance problems and integrating different jurisdictions, institutions and agreements. Is this realistic or desirable?

Tait (1992, p.34) has argued the following about values and interests:

> It is important to understand the extent to which interests and values are involved in any conflict, because tactics that will improve one will make the other worse... A conflict of interest can sometimes be resolved by simply giving more or better information to change the understanding of the potential impact... Where there is... genuine

divergence of interests, the various parties can bargain with one another until a satisfactory settlement is reached. Conflicts of value, on the other hand, can be exacerbated by both tactics. Protagonists in a value conflict will only accept information that is in accordance with their beliefs; everything else will be treated as propaganda and its source discredited. Attempts at bargaining to reach a settlement will be treated as bribery, trading principles for cash – again leading to a worsening of conflict.

This is a useful starting point for our analysis because it encourages us to look more closely at the values and interests that are implicit and explicit in international agreements and institutions, and in the actions of governments and others who interact with them as part of an international governance process.

As discussed above, Isaac (2002) has argued that all international agreements have an economic and social objective. Their economic objective is to improve the efficiency of the market system. Their social objective is to ensure that market activity takes place in a way that is consistent with the preferences and expectations that exist within jurisdictions. These objectives can in turn be linked to different understandings of how the world should be integrated:

> International economic integration adopts economic principles to explain appropriate regulatory development and regulatory integration strategies for a jurisdiction. Accordingly, rational, optimizing behaviour fulfilling the economic function governs policy development, shapes social regulations and forms the basis for dealing with social regulatory barriers through trade diplomacy. Alternatively, international social integration adopts a perspective that argues that markets are embedded in normative constructs so that the economic perspective is meaningless if separated from social realities. (Isaac, 2002, p.13)

As an analytical starting point these ideas can help us to understand important differences between the WTO, the Codex Alimentarius Commission and the UN Convention on Biological Diversity.

Although all the agreements and institutions discussed above have economic and social objectives built into them, the relative importance of one or the other is clearly different in each case. The primary aim of the WTO agreements, for example, is the achievement of economic integration and efficient market activity. It is clear that the economic objective and economic rationality dominate in this case. The purpose of the Codex Alimentarius, on the other hand, is arguably only the facilitation of economic integration via the establishment of global food standards. This suggests some balancing of economic and social objectives. Finally, and in relative terms revealing the greatest influence of social objectives, the Biosafety Protocol aims to shape economic integration *ex ante* to create a global (and normative) social construct. This approach, therefore, provides a useful perspective on the different values and interests that are embedded in the agreements and institutions.

The work of Isaac (2002) is also useful because he extends his use of the concepts of economic and social rationalities beyond the categorisation of international agreements and institutions. He also uses these ideas to explain how agreements and institutions of different types are likely to treat new technologies – e.g. biotechnology. He describes the link between economic and social rationalities and new technologies as follows:

> [The] economic perspective generally assumes that technology and innovation are vital factors of economic growth and welfare. As a result, it supports a regulatory framework that encourages technological progress. For instance, it is quite common for economic analysis to support "scientific-rationality" approaches to regulating the risk of new technology... The economic- and scientific-rationality perspectives are similar, in that they decompose complex behaviour and actions into causal-consequence models, which are then used to forecast outcomes. (Isaac, 2002 pp.16-17)

> [The] "social-rationality" approach holds that it is insufficient to view new technology and innovations simply as a positive force in economic growth. Instead, the social implications of science must be considered and, under this consideration, new technology may not always be greeted without reservation – despite its potential to improve economic growth. (Isaac, 2002 p.21)

These comments draw attention to the levels of enthusiasm and scepticism regarding new technologies in international agreements and institutions and how these are linked to the economic and social rationalities that underpin them. Can these ideas also help us to understand how international agreements and institutions use science to control new technologies?

The evidence above clearly shows that the SPS Agreement, the standards of the Codex Alimentarius and the Biosafety Protocol treat the problems of uncertainty and ignorance in risk assessment differently. At one extreme the SPS Agreement only acknowledges uncertainty as a relatively short-lived problem, and one that only justifies interim measures by governments. At the other extreme the Biosafety Protocol includes the precautionary principle and thereby acknowledges more uncertainty and establishes a different standard. Such differences can be understood more easily if the values that underpin these agreements are kept in mind, along with the associated view of new technology. The limited treatment of uncertainty in the SPS Agreement is, in part, explained by the enthusiasm for new technology in the economic rationality that underpins the agreements of the WTO. Put simply, a limited treatment of uncertainty leaves less room to restrict a technology on the grounds of risk. In reverse, the less circumscribed treatment of uncertainty in the Biosafety Protocol can be linked to the more circumspect position on new technology that is a part of the social rationality that underpins this agreement. The more developed treatment of uncertainty results in more scope to restrict the technology on the grounds of risk.

Conclusion – Some Reflections on Governance and Integration

What are the implications for "governance" and "integration"? There are currently a number of overlapping and contradictory international agreements and institutions. It is not clear which is the most authoritative or how they relate to each other. Each one has a part to play in the regulation of GMOs at the global level. In this case science does not offer a way of integrating them (or of overcoming related disagreements), partly because it is itself shaped by the values and interests involved.

The GMO case is complicated further by the fact that the two largest trading blocs in the world are in conflict with each other. The US, by launching its formal complaint at the WTO, aligned itself with this regime. The EU, in contrast, moved quickly to ratify the Biosafety Protocol, which the US will not ratify. Other countries, to a large extent, are remaining neutral or lining up behind one or the other.

From one perspective – to some extent a managerialist one – the lack of coherence in international agreements and institutions is problematic (and part of "the problem"). However, from another perspective – largely a political one – it is realistic, useful and perhaps even essential. Because of the lack of integration the system has a built-in ability to cope, to some extent, with the strain that is being imposed on it by the US-EU conflict. The formal complaint by the US at the WTO is problematic because it risks forcing the system to clarify things that were most useful when they were ambiguous and unclear.

This chapter suggests that there are considerable risks associated with assuming that "integration" is a good thing. More generally it suggests that there might be problems associated with linking the ideas of "governance" and "integration". Governance is a complex process involving the interaction of multiple stakeholders, often with different definitions of "the problem", in numerous fora at different political levels. It seems unlikely that this is compatible, practically and theoretically, with the idea of integration.

Acknowledgement

The authors would like to acknowledge the financial support of the UK's Economic and Social Research Council (ESRC). This chapter draws on a number of research projects funded by the ESRC. Valuable comments on earlier versions were made by a number of colleagues in The Open University and the University of Edinburgh.

A version of this chapter was presented at the Goodenough College *Trading Genes: the Power of the Market in Shaping a New Genomic Order* conference, London, 27 March 2003. It was also submitted as written evidence to the Science and Technology Select Committee of the House of Lords (Sub-Committee I) in October 2003 as part of their investigation into the role of science in international agreements.

References

Afonso, M. (2002), 'The relationship with other international agreements', in C. Bail, R. Falkner and H. Marquard (eds), *The Cartagena Protocol on Biosafety*, pp.423-437, London: RIIA/Earthscan.

Andrén, R. and Parish, B. (2002), 'Risk Assessment', in C. Bail, R. Falkner and H. Marquard (eds), *The Cartagena Protocol on Biosafety*, pp.329-337, London: RIIA/Earthscan.

Bail, C., Falkner, R. and Marquard, H. (eds) (2002), *The Cartagena Protocol on Biosafety*, London: RIIA/Earthscan.

CAC – Codex Alimentarius Commission (1997), 'Report of the Twenty-Second Session of the Codex Alimentarius Commission', Geneva, 23-28 June 1997, ALINORM 97/37.

CAC – Codex Alimentarius Commission (1999), 'Report of the Twenty-Third Session of the Codex Alimentarius Commission', Rome, 28 June – 3 July 1999, ALINORM 99/37.

CAC – Codex Alimentarius Commission (2001a), 'Report of the Fifteenth Session of the Codex Committee on General Principles', Paris, 10-14 April 2000, ALINORM 01/33.

CAC – Codex Alimentarius Commission (2001b), 'Report of the Sixteenth Session of the Codex Committee on General Principles', Paris, 23-27 April 2001, ALINORM 01/33A.

CAC – Codex Alimentarius Commission (2001c), 'Report of the Twenty-Fourth Session of the Codex Alimentarius Commission', Geneva, 2-7 July 2001, ALINORM 01/41.

CAC – Codex Alimentarius Commission (2001d), 'Report of the Second Session of the Codex Ad Hoc Intergovernmental Task Force on Foods Derived from Biotechnology', Chiba, 25-29 March 2001, ALINORM 01/34A.

CAC – Codex Alimentarius Commission (2003a), 'Report of the Twenty-Sixth Session of the Codex Alimentarius Commission', Rome, 30 June – 7 July 2003, ALINORM 03/41.

CAC – Codex Alimentarius Commission (2003b), 'Report of the Third Session of the Codex Ad Hoc Intergovernmental Task Force on Foods Derived from Biotechnology', Yokohama, 4-8 March 2002, ALINORM 03/34.

Dratwa, J. (2002), 'Taking risks with the precautionary principle', *Journal of Environmental Policy and Planning*, **4**, pp.197-213.

FAO/WHO – Food and Agriculture Organization/World Health Organization (2000), 'Safety Aspects of Genetically Modified Foods of Plant Origin', Report of a Joint FAO/WHO Expert Consultation on Foods Derived from Biotechnology, 29 May – 2 June.

FoEE – Friends of the Earth Europe (1999), FoEE Biotech Mailout, **5**(5), 31 July 1999.

Fukuyama, F. (1992), *The End of History and The Last Man*, Harmondsworth: Penguin.

Gore, C. (2000), 'The Rise and Fall of the Washington Consensus as a Paradigm for Developing Countries', *World Development*, **28**(5), May.

Graff, L. (2002), 'The precautionary principle', in C. Bail, R. Falkner and H. Marquard (eds), *The Cartagena Protocol on Biosafety*, pp.410-422, London: RIIA/Earthscan.

Harriss, J. (2000), 'The Second Great Transformation? Capitalism at the end of the twentieth century', in T. Allen and A. Thomas (eds), *Poverty and Development in the 21st Century*, Oxford: Oxford University Press.

Isaac, G. (2002), Agricultural Biotechnology and Transatlantic Trade: Regulatory Barriers to GM Crops, Oxford: CABI Publishing.

Jasanoff, S. (1993), 'Bridging the two cultures of risk analysis', *Risk Analysis*, **13**(2), pp.123-29.

Kimbrell, E. (2000), 'What is Codex Alimentarius?', *AgBioForum*, **3**(4), pp.197-202.

Pew, (2003), U.S. vs EU: An Examination of the Trade Issues Surrounding Genetically Modified Food, Pew Initiative on Food and Biotechnology, August 2003.

Polanyi, K. (1944, 1957), The Great Transformation: The Political and Economic Origins of Our Time, Boston: Beacon Press.

Scott, J. (2000), 'On Kith and Kine (and Crustaceans): Trade and Environment in the EU and the WTO', in J. Weiler (ed.), *The EU, the WTO and NAFTA: Towards a Common Law of International Trade*, Oxford University Press, Oxford, pp.125-167.

Stiglitz, J. (2002), *Globalization and its Discontents*, London: Penguin.

Tait, J. (1992), 'Biotechnology – Interactions between Technology, Environment and Society', FAST Programme: Biosphere and Economics. Brussels.

Chapter 10

A New Mode of Governance for Science, Technology, Risk and the Environment?

Joyce Tait and Catherine Lyall

Why Do We Need New Policy Approaches?

The new governance agendas that are the subject of this book have not arisen merely because of some collective whim on the part of bored politicians or policy-makers seeking change for its own sake. They are being driven by a greater perceived need for policy integration reinforced by experience of policy failures when complex issues are constrained within too narrow a policy boundary.

This book has considered a range of policy-relevant problems and how policy-makers are approaching them (or in some cases failing to do so). The first three chapters outline the new approaches to governance themselves; the challenges that have led to them; the changes they are making; and how they are being developed. Subsequent chapters outline experience of their application in a wide range of areas – life sciences, risk analysis, rural policy, transport, environmental policy and international trade agreements.

Our perspective is generally to consider the processes of policy-making – what problems do policy-makers perceive, how are they attempting to make progress in resolving them and how effective are the resulting governance systems? Elsewhere, we have considered similar questions, particularly in the context of the life sciences, from the perspective of the policy targets, the organisations and individuals whose businesses and behaviour policy-makers are attempting to influence (Tait et al., 2004). In that case, the focus is on "what works" for the policy targets rather than "what works" for the policy-makers. Overall, any effective set of policies will need to strike a balance between both these aspects.

Globalisation has been one of the main drivers of these new policy approaches. The increasingly inclusive, globally-based organisation of scientific research and the truly global reach of the companies that contribute both to scientific discoveries and to the commercial exploitation of the products of science places new demands on policy-makers seeking to promote or to regulate these activities, as discussed by Murphy and Chataway in Chapter 9.

Related to globalisation, the rapidity of technological change is presenting major challenges to policy-makers, both in the promotion of science and innovation and in its management and control (see Spinardi and Williams, Chapter

3 and Reiss and Tait, Chapter 4). Medical developments in life sciences, for example in stem cells and genetic databanks, are evolving much more rapidly than the policy and regulatory systems that have traditionally governed these sciences. New types of product are also crossing the boundaries of traditional regulatory systems. For example, how should we regulate a food crop that has been genetically modified to be herbicide- and insect-resistant and also to produce chemicals that will protect against, for example, heart disease or cancer – as a pesticide, a food supplement or by a new regulatory process designed to deal with all genetically modified crops as an entirely new regulatory category? Disagreements on this specific point underlie some aspects of the very long-running dispute between the USA and Europe over the regulation of GM crops (Tait and Levidow, 1992; Murphy and Chataway, Chapter 9).

In another context altogether, that of rural land use, technological change is opening up new ranges of opportunities that may lead to more effective policies for sustainable development of rural communities (Rennie, Chapter 6).

Globalisation and rapid rates of technological change are together leading to new types of systemic problem where new technology allied to new commercial pressures and linked to age-old human desires, for example for mobility, are leading to outcomes where all will be losers but no single group of actors, stakeholders or regulators has the incentive or the ability to effect change (Adams, Chapter 7). Also included in this systemic category are modern, international information networks and electricity grids where there have been recent catastrophic failures affecting hundreds of thousands of people often in more than one country. So far, neither old-style govern*ment* nor new style govern*ance* has found an answer to these "wicked" problems (6, 1977).

Policy-makers themselves have recognised some of these problems and in a few cases have attempted to develop more general integrated approaches, for example to risk-related issues (McQuaid, Chapter 5).

In relation to science and technology, a complicating factor for the introduction of new governance agendas is the need at the same time to retain significant elements of old-style government in the form of straightforward top-down regulation (see Hertin and Berkhout, Chapter 8). Command and control, backed up by sanctions and penalties will continue to be needed to regulate the safety to human health and the environment of innovative products and of the processes used to develop them. Perri 6 (Chapter 2) discusses the available regulatory instruments under the headings of "control", "inducement" and "influence", and Reiss and Tait (Chapter 4) and Spinardi and Williams (Chapter 3) focus on the mechanisms for the responsible promotion of science and technology.

As Perri 6 notes (Chapter 2), the science and technology-related issues which are most contentious are those around which economic, political and social change are organised, changing the allocation of resources between groups. Technological change can be seen in terms of (i) risks that need to be changed or managed, (ii) opportunities for financial gain or improvements in health, or (iii) societal challenges arising from the decline of established industries. Any innovation worthy of attention from a governance perspective is likely to have all three types of effect, impacting selectively on different societal groups and acting as a potential trigger for conflict.

Thus, many of the complicating factors identified in the application of the governance agenda to science and technology-related issues arise from complex interactions between the still-necessary, govern*ment*-based regulation and control and the main themes of the govern*ance* agenda, particularly the contrasting and sometimes incompatible requirements for policy decisions to be evidence-based and at the same time for a greater degree of stakeholder engagement in policy decision-making. These are some of the challenges that are faced in attempting to foster policy integration.

Promotion and Regulation of Science and Innovation

The promotion of science and innovation takes a variety of forms, ranging from the funding of "blue skies" scientific research, through various processes of development of technology-based products, to the opening up of markets for innovative products. Policy-makers have a role in all these processes, in setting up programmes and priorities for public funding of science, in creating a supportive environment for commercialisation of research (for example, in encouraging the setting up of small companies and ensuring effective regimes for the protection of intellectual property), and in managing the market conditions for the introduction of new products. The chapters of this book provide numerous examples of such policy initiatives in action.

Reiss and Tait (Chapter 4) discuss the evolving role of Foresight as a policy instrument to guide the prioritisation of public support for particular areas of innovation and, where such support has been provided for the life sciences, present the results of research undertaken to gauge its effectiveness. Spinardi and Williams (Chapter 3) give a comprehensive overview of research on the challenges facing policy-makers in understanding the processes involved in promoting "breakthrough" science and technology and also the possibilities for, and limitations of, policy in governing the subsequent innovation trajectory.

Internationally, as Perri 6 points out (Chapter 2), the World Trade Organization (WTO) has had a big impact on the governance of opportunity for innovative products to gain access to markets by encouraging the free movement of capital, labour, services and goods. Perri 6 also reflects on the fears that deregulation, driven by global competition, will lead to a "race to the bottom" where businesses move to the least regulated states. There is some evidence of this happening, for example, in the falling investment in GM crop research and development in Europe and the movement of companies to the United Sates (Reiss and Tait, Chapter 4), although this has more to do with prolonged uncertainty about the final nature of the European regulatory system than with its actual stringency. There are also examples where international agreements have led to a global tightening of standards, a "race to the top" (Tait and Bruce, 2004).

Also in the international context, Hertin and Berkhout (Chapter 8) refer to the EU Lisbon Agenda which aims to make Europe's economy "the most competitive and dynamic knowledge-based economy in the world, capable of sustainable economic growth with more and better jobs and greater social cohesion". However, they also note that, from the regulatory perspective, it was not until the Gothenburg European Council a year later that an environmental dimension was added to this strategic policy initiative. Doubts about the political will to give greater emphasis to environmental objectives were also raised by several aspects of the implementation of the EU Sustainable Development Strategy which was seen to have an additive rather than integrative character, failing to stimulate public participation.

This is a common feature of many policy agendas where promotion of science, technology and innovation precedes, sometimes with a considerable time lag, consideration of potential risks to people or the environment. This failure simultaneously to consider both the benefits and costs of innovation underlies much European public pressure for greater implementation of the precautionary principle in the governance of innovation.

In many cases, technological innovation can provide an answer to environmental problems caused by earlier generation products. For example, modern vehicles are considerably more efficient and less polluting than their predecessors and these changes have been stimulated largely by creative use of policy and regulatory instruments to drive innovation in this direction. However, as Adams (Chapter 7) points out, in the case of the hypermobility problems he identifies, the promotion of technology as a solution to transport-related pollution has actually compounded the overall problem.

The Role of Evidence in Policy-making

One of the key requirements of the new governance approach is that policies should be evidence based, in several senses of the term. They should be based on evidence that policy intervention is required in a particular area, the specific policy instruments adopted and the threshold levels for regulatory standards should be based on evidence that they are likely to work as intended, and monitoring should be undertaken to check that they are indeed working as intended. These points are reflected in the focus of modern governance on "what works".

As the degree of stakeholder and public engagement in policy development increases, the evidence base for new policy initiatives is more likely to be contested by at least some stakeholder interests, increasing the timescale and the cost of policy-making. Also, as Spinardi and Williams (Chapter 3) and Reiss and Tait (Chapter 4) demonstrate, obtaining unequivocal evidence for the effectiveness of a policy initiative (i.e. deciding whether it works or not) is difficult to achieve using evidence that does not have exorbitant collection costs and that can be gathered on a timescale that is useful to policy-makers.

To give some examples of these points, Murphy and Chataway (Chapter 9) describe the example of the Codex Alimentarius Commission's approach to the potential allergenicity of GM crops, involving arguments for and against a "decision tree" approach or a "preponderance of evidence" approach. As they point out, the explanation for these debates goes beyond questions of scientific evidence, and brings in the values and interests of those involved. The preponderance of evidence approach may make it easier for a GM food to pass a risk assessment because no single test that raises a doubt about allergenicity will result in its being rejected, partly explaining why pro-GM stakeholders supported this approach. The decision tree approach is more demanding and was supported by stakeholders who take a more cautious or sceptical approach to the technology.

Rennie (Chapter 6) discusses the difficulty of finding indicators by which progress towards sustainability can be measured. The problem is not a lack of indicators per se but a lack of agreement on which indicators are the most appropriate for any given sector, scale, and geographical region and most indicators lack any universal applicability. In order to develop policies to improve the effectiveness of governance, it is necessary to have a consensus on what we want to move towards. Until recently, sustainable development was regarded as synonymous with the maintenance of environmental conditions and habitats but, under the new governance agenda, the concept of sustainable development has moved towards a more inclusive approach which also brings in concerns about social, economic and political sustainability (Tait and Morris, 2000).

Progress with Impact Assessment in the EU is described by Hertin and Berkhout (Chapter 8) as a possible instrument to encourage the use of evidence in an integrative manner in policy-making. However, evidence so far on the implementation of this policy device is not encouraging – many reports only consider one policy option, are narrowly focused on direct economic impacts and explore social and environmental impacts only briefly, some are not evidence-based, and few are transparent about how and by whom the assessment has been carried out.

McQuaid (Chapter 5) offers a number of useful suggestions on gathering evidence for the effectiveness of a policy initiative like the Inter-departmental Liaison Group on Risk Assessment (ILGRA). The qualitative criteria suggested for the operation of ILGRA itself include style, relevance, self adaptation, learning, innovation and transparency. Evidence is also given in this chapter of several clear impacts on policy. However, such cross departmental government initiatives often rely heavily on the commitment of one or two key individuals and it is not yet clear whether the approach embodied in ILGRA has become firmly embedded in the UK policy lexicon.

The examples given in the preceding chapters indicate a clear commitment by policy-makers to a stronger evidence base for planning and monitoring the policies they are developing. However, partly because of the long timescales involved in planning and implementing policies for science, technology and innovation, it will be some time before we have evidence that this component of the new governance mode is working.

Stakeholder Engagement in the New Governance Agenda

Where science and technology-related policy issues have generated conflict and become intractable, the solution offered as part of the old-style government approach was greater public understanding of science achieved by targeted education campaigns. This "deficit model" which assumes that better public understanding of science will result in greater public approval of technology has been very effectively challenged by social scientists and its defects have also been acknowledged by the Royal Society (1997) and the House of Lords Select Committee on Science and Technology (2000). However, this model is still implicit in the actions and reports of many science and policy communities (for example, Scottish Executive, 2001).

In contrast, new governance modes advocate an increasing role for non-governmental actors and stakeholders in policy-making, generally favouring public rather than private or commercial groups, as a means to mitigate public controversy over new technology developments. As Lyall and Tait note in the first chapter, this is leading to increasingly complex state/society relationships where networks rather than hierarchies dominate the policy-making process and there is a blurring of public/private boundaries (Bache, 2003).

Reiss and Tait (Chapter 4) refer to the discrediting of the "picking winners" approach to Foresight and its replacement by a more inclusive approach which engages with a wider range of stakeholders, including public representatives. In the past, the acknowledged role of the public has been as consumers, using their purchasing decisions to pick the winners from the range of science and technology offerings which eventually reach the market place. However, particularly among European public interest pressure groups, there is now increasing dissatisfaction with this reliance on the market place as a discriminator among products, again reinforcing demands for greater public involvement in policy-making for science and innovation from the earliest stages in the innovation process (Wilsdon and Willis, 2004).

As Rennie also points out in Chapter 6, rural policy and practice has gone beyond being of interest only to farmers and includes environmentalists, countryside ramblers, other outdoor recreational interests, heritage societies, town-dwellers with views on the countryside, urban shoppers who want to trace the origins of their supermarket purchases, specialist growers, organic food enthusiasts, and protesters against genetically modified food. This spreading constituency presents opportunities for policy-makers to bring rural issues more centre-stage in governance but also creates problems in establishing a common vision for the countryside. With so many overlapping and contradictory interest groups, set against a constantly shifting policy regime, it is becoming harder to establish a policy consensus on the overall objectives.

Offering a different perspective on stakeholder engagement, Adams (Chapter 7) points out that hypermobility is likely to lead to much weaker *local* engagement of stakeholders. The corollary of this is that a broader range of stakeholders is more likely to engage with policy issues at a national or international level and, because the issues involved are more remote, to engage on the basis of values or ideology rather than local personal interest. In such cases, conflict and polarisation of the issue is likely to arise, resolution of the conflict will be more difficult to achieve, and the evidence base for policy decisions will be increasingly challenged (Tait, 2001).

Indeed, there are unresolved, and generally unacknowledged, tensions in the expectations that, through new governance approaches, policy-makers will simultaneously engage with a wider range of stakeholders and also increasingly base their decisions on evidence. A common tactic among the diverse groups and networks of stakeholders that tend to engage with policy decisions on STI issues is to challenge data put forward as evidence and for each "side" in a controversy to give selective attention to evidence that fits with their own objectives. Such challenges downgrade the value of research findings as evidence to support decision-making, and policy-makers who attempt to take decisions on the basis of available evidence are finding that science and technology are becoming increasingly ungovernable as the evidence base for decisions is challenged and eroded.

These issues are closely related to the uncertainty that often surrounds the evidence used as a basis for policy decision-making. Under previous policy approaches, uncertainty was often resolved after a product had been approved and on the market for a period of time. In the case of pesticides, for example, defects that emerged after a product had been in use for some time were dealt with by banning the product and replacing it with an alternative or by altering its formulation or pattern of use (Tait and Levidow, 1992). This allowed a great deal of scientific, organisational and policy learning about products, their regulation and use. However, under the governance agenda, public stakeholder groups increasingly invoke the precautionary principle in an attempt to ensure that defective products do not even reach the marketplace, as discussed by Murphy and Chataway (Chapter 9). Although this is a laudable objective, it could not be described as evidence based.

Another issue that is addressed briefly by Rennie (Chapter 6) is that of consultation fatigue among stakeholders. His references to being buried by bureaucracy and issues suffering death or petrification by serial consultation reflect current experience in several social policy areas, although perhaps not yet in the context of science and technology.

The Feasibility and Desirability of Policy Integration

Pressures for Integration

A consistent theme throughout the new mode of governance is the need for more integrated or "joined up" policy approaches to remove contradictions, inconsistencies and inefficiencies caused when policies or regulations emerging from different government departments or different levels of government (regional, national, international) contradict one another or provide incompatible or contradictory signals to policy targets. Policy integration is also needed to deal with the complexity and uncertainty associated with many decisions concerning science and technology. However, integration has itself become more difficult as the diversity and policy competence of interested stakeholders and publics has increased.

At least some of the pressure towards policy integration arises from policy-makers themselves. Risk is a good example of an issue that pervades policies across a wide range of government departments and for which there is a recognised need for integration of policy-making. As McQuaid (Chapter 5) points out, managing risks solely on the basis of probabilistic estimates of physical harm is unlikely to succeed. Public values and perceptions of risk must be integrated into the decision-making process, although the extent to which this is possible depends, for example, on constraints such as the need to meet EU obligations or their legitimacy if they conflict with principles of fairness and equity. There is also a need to balance the extent to which risks are prevented or controlled against the resources required and the economic benefits foregone.

Likewise, the EU sustainable development strategy described by Hertin and Berkhout (Chapter 8) puts as much emphasis on processes of governance as it does on substance and emphasises the need for a new policy approach. However, despite these undoubted pressures, given the complex and often contentious nature of the underlying issues, policy integration is not proving to be straightforward or to be a "quick fix".

Mechanisms for Integration

Various mechanisms for policy integration are proposed or described in the chapters of this book. Lyall and Tait (Chapter 1) describe the pressures for co-ordination of policy processes across functional boundaries, both vertical and horizontal, two-way and one-way, with the observation that vertical integration is easier to achieve in practice than horizontal.

However, even vertical integration is not without its problems. At EU level, for example, environmental policy integration is dependent on better integration in member states, largely because they hold much of the power in EU policy-making and are also responsible for implementation. Unless national governments improve their internal coherence, the political capacity of the European Commission to promote integration should not be over estimated (Hertin and Berkhout, Chapter 8).

Despite these documented difficulties, McQuaid (Chapter 5) describes an example of horizontal integration across a range of UK government departments by means of a very effective policy network in the context of risk regulation, ILGRA. On the other hand, Lyall and Tait (Chapter 1) point to the rather weak mechanisms available in the UK for policy integration for science, technology and innovation, taking place through cross-departmental membership of committees, rather than the more genuinely holistic governance approach referred to by Perri 6 in Chapter 2.

Although ILGRA achieved some notable successes in the area of risk regulation, McQuaid (Chapter 5) also points to the problem of demonstrating this effectiveness, of providing evidence for it. Qualitative, rather than quantitative, methods were seen to be most appropriate, and key questions for discussion were: to what extent was ILGRA successful in the process of joining up policy rather than being just another government co-ordinating committee; what actual impact did it have in the period under review; and where does it now stand in the overall scheme of things?

Fostering policy networks is only one part of the overall process of policy integration. ILGRA, for example, also identified the need for the development of frameworks for integrating risk estimates, public perceptions, the need for trade offs and other factors in the decision-making process, and the steps that departments and regulators could take to gain acceptance of their frameworks and of the consequent decisions as representing the best trade offs in particular situations. An important part of this approach was the development of a common framework which included the active engagement of stakeholders in all stages so that they can influence the assumptions and value judgements that permeate the procedure and hence concur more readily with decisions arising from it.

According also to Hertin and Berkhout (Chapter 8), there are many ways in which policy approaches to science, technology, risk and the environment could be better co-ordinated and aligned across different vertical levels of policy-making, between different governmental and non-governmental actors and throughout the process of technological innovation. Taking as an example integration (or its absence) between environmental policy and key sectoral policy areas such as transport, planning, agriculture and energy, they point to the overriding concern of energy departments to supply cheap and secure energy, rather than to minimise air pollution or to prevent climate change. Adams (Chapter 7) observes the same kinds of relationship between transport planners and the policies of other government departments. There is a belief that better decisions would arise from institutional arrangements that established a competitive relationship between promoting and regulating departments, providing a contest of ideas that is more likely to deliver clarity and to improve the rationale and evidence base for a given policy action. Key elements of this approach are sectoral agenda setting, horizontal communication, sectoral capacity building and policy learning.

Difficulties Experienced

Undertaking genuine policy integration takes more time and usually costs more money than more compartmentalised decision-making and the benefits may not accrue to the policy department that bears the cost. It is thus important but, as we have noted, difficult to produce evidence of the effectiveness of any particular integrated policy initiative.

Among the most serious difficulties encountered, Hertin and Berkhout (Chapter 8) point to the bias against integrated technological responses in some policy communities. In order to minimise interference in their affairs, sectoral departments have an interest in reducing the involvement of environmental administrations in the process of policy formulation. This tends to delay and diminish the effectiveness of inputs from environmental interests. These relationships between intra-governmental and policy outcomes can also be self-reinforcing and they point to the self preservation instincts of institutions established to administer some pollution control policies. Even where there is no active resistance to integration on the part of policy-makers, there are more subtle career-related disincentives for policy-makers to stray across functional or departmental boundaries.

In the case of risk-related issues, McQuaid (Chapter 5) points out the inherent difficulty in seeking to achieve integration by the pursuit of evidence-based policies, particularly for contentious issues arising from the collision of scientific developments and public unease about them. As risk is itself a representation of uncertainty about causes and consequences of activity, there will be conflicting judgements about the significance of a particular risk which will need to be countered by dialogue involving a range of actors and stakeholders inside and outside government. In addition, the criteria and value judgements that influence the evaluation of risk must inevitably figure in the process of dialogue.

Another unrelated difficulty is the continuing need for consistency and continuity in regulatory systems and the greater global inter-connectedness so that there is less freedom for national governments to develop their own governance approaches.

Conclusion

In attempting to make progress towards integrated policy approaches, there are few clear and evident success stories. For the complex, systemic contexts where policy integration is a relevant aim, change can never be straightforward. We shall not understand how the governance of technology works or how it could work differently if we do not see the whole system (Perri 6, Chapter 2). The closest we get to a success story in terms of the new governance agenda, among the chapters included in this book, is ILGRA and even here, as McQuaid (Chapter 5) admits, an independent outside observer may be more critical than he is.

Adams (Chapter 7) notes that there are no mechanisms to integrate the social impacts of hypermobility into planning and transport policies. Likewise, Hertin and Berkhout (Chapter 8) note that the EU sustainable development strategy does not provide the necessary leadership and it has not strengthened civil society engagement.

One of the biggest challenges in integrating the various elements of the new governance agenda is the potential incompatibility between evidence-based decision-making and greater stakeholder engagement. Policy decisions that are ostensibly science-based have always been influenced by interests and values. Wider stakeholder and public engagement inevitably leads to demands to bring a broader range of interests and values, often expressed more forcibly, into consideration in these decisions. So far we have not learned to integrate stakeholder engagement in issues related to science, technology, risk and environment in ways that do not undermine the role of evidence in decision-making or indeed make it redundant.

At the EU level, the emphasis by Hertin and Berkhout (Chapter 8) on the integration principle in key policy documents contrasts with a lack of progress in integration in EU policy-making. EU institutions have traditionally functioned in a hierarchical and segmented fashion and the different sectoral formations of the Commission work largely independently. The relationship between DG Environment and sectoral Directorates-General (DG Enterprise and DG Agriculture) involves deeply entrenched conflicts of interest, cultural differences and controversies which prevent the re-alignment of policy objectives and strategies.

Also in the international context, Murphy and Chataway (Chapter 9) found very few instances where the new agendas have resulted in better governance. However, they do not see this as necessarily a bad thing and point out that lack of formal integration can leave the way open to creative, informal initiatives that might have been constrained by a more tightly integrated system. They make the point that, from a managerialist perspective, the lack of coherence in international agreements and institutions is problematic, and indeed it is part of "the problem". However, from another perspective – largely a political one – lack of integration is pragmatic, useful and perhaps even essential. The formal complaint by the US at the WTO that is the subject of Chapter 9 can be seen as problematic because it risks forcing the system to clarify things that were most useful when they were ambiguous. In this case, there are risks associated with assuming that integration is a good thing and indeed there may even be problems with the very concept of linking the ideas of "governance" and "integration".

Thus, integration will always be partial; and complete integration is neither feasible nor desirable. Integration itself is a neutral concept, neither good nor bad: its advantages depend on how it is used, by whom, and in what policy context. Indeed, in the hands of a despot or a politically powerful interest group it could be used to essentially undemocratic ends.

The social sciences have been very active and highly effective in exposing the defects of the govern*ment* agenda of the 1980s and before. However, social science research has so far been less effective in exposing these tensions in the new govern*ance* agenda. What the governance approach debated in this book has done is turn the spotlight on the need to develop strategies and procedures to help decision-makers to govern science, technology, risk and the environment in a way that makes best use of appropriate systemic analyses on the basis of the best available evidence from both social *and* natural sciences.

References

6, P. (1977), *Holistic Government*, London: Demos.

Bache, I. (2003), "Governing through Governance: Education Policy Control under New Labour", *Political Studies*, **51**(2), pp.300-314.

House of Lords Select Committee on Science and Technology (2000), *Science and Society*, London: Stationery Office, HL 38/38-I.

Royal Society (1997), *Science, Policy and Risk*, London: The Royal Society, p.87.

Scottish Executive (2001), *A Science Strategy for Scotland*, Edinburgh: The Stationery Office.

Tait, J. and Levidow, L. (1992), 'Proactive and Reactive Approaches to Risk Regulation: the Case of Biotechnology', *Futures*, April, pp.219-231.

Tait, J. and Morris, D. (2000), 'Sustainable Development of Agricultural Systems: Competing Objectives and Critical Limits', *Futures*, **32**, 247-260.

Tait, J. (2001), 'More Faust than Frankenstein: the European Debate about Risk Regulation for Genetically Modified Crops', *Journal of Risk Research*, **4**(2), 175-189.

Tait, J., Chataway, C. and Wield, D. (2004), *Governance, Policy and Industry Strategies: Agro-biotechnology and Pharmaceuticals*, Innogen Working Paper 12, www.innogen.ac.uk.

Tait, J. and Bruce, A. (2004), 'Global Change and Transboundary Risks', in McDaniels, T. and Small, M. (eds), *Risk Analysis and Society: an Interdisciplinary Characterisation of the Field*, Cambridge: Cambridge University Press, pp.367-419.

Wilsdon, J. and Willis, R. (2004), *See-through Science: Why Public Engagement Needs to Move Upstream*, London: Demos.

Index